THE STAFF DESIGNER
GROW, INFLUENCE, AND LEAD AS AN INDIVIDUAL CONTRIBUTOR

Catt Small

NEW YORK   2025

"The perfect blueprint for making real impact as an independent contributor."

—Ridd (Michael Riddering)
Co-founder of Inflight and host of the *Dive Club* podcast

"Staring into the senior void? Catt's book is like a friendly coach, giving you cross-company insights and frameworks to drive your UX career forward."

—Cheryl Platz
Former Principal UX Designer at Microsoft and
the Gates Foundation;
author of *Design Beyond Devices* and
The Game Development Strategy Guide

"This book should be required reading for designers who start to wonder what's next. It takes the gnarliest challenges of climbing the individual contribution ladder and presents bite-sized, approachable frameworks and lessons anyone could implement the same day."

—Abby Covert
Chief Sensemaker at The Sensemakers Club

"Staff designers are more important than ever in today's environment. Catt's book will help you thrive as one."

—Jenny Wen
Design Lead at Anthropic

"A ridiculously practical guide to an often-undefined area of super senior leveling and success."

—Hang Xu
Design Recruiter and founder of Designer Friends

"In the fast-paced, chaotic world of start-ups, many designers grow in seniority out of necessity, rather than readiness. Catt's book offers actionable insights from a diverse range of experiences and voices, with thoughtful research that I found easy to learn from and apply in my own work, especially as I recently assumed the mantle of staff designer. I highly recommend this book to anyone on their journey to seniority, and beyond."

—Asia Hoe
Staff Product Designer and public speaker

"An essential read for any designer looking to grow beyond the senior level with influence and impact."

—Femke van Schoonhoven
Design Manager, femke.design

"Catt's book nails the mindset shift every staff designer has to go through: navigating ambiguity, influencing the business, and scaling your impact. An essential read for every designer serious about leveling up."

—Senongo Akpem
VP of Design at Nava PBC

The Staff Designer
Grow, Influence, and Lead as an Individual Contributor
By Catt Small

Rosenfeld Media, LLC
125 Maiden Lane
New York, New York 10038
USA

On the Web: www.rosenfeldmedia.com

Please send errata to: errata@rosenfeldmedia.com

Publisher: Louis Rosenfeld

Managing Editor: Marta Justak

Interior Layout: Danielle Foster

Cover Design: Heads of State

Illustrator: Catt Small

Indexer: Marilyn Augst

Proofreader: Sue Boshers

© 2025 Catt Small

All Rights Reserved

ISBN: 1-959029-77-0

ISBN 13: 978-1-959029-77-9

LCCN: 2025936429

To Mom and Dad for always being in my corner,
and to my incredible grandparents each of whom
inspired me with their stories.

HOW TO USE THIS BOOK

Who Should Read This Book?

This book was written for UX practitioners who are considering the meaning of the staff role in different ways. If you are one of the following, you should definitely read this book:

- Staff designers in need of new tactics to shift from reactive to proactive, build influence, and become the leader they want to be.
- Senior designers who want to get promoted to the staff level by identifying and filling their competency gaps.
- Senior UX practitioners who want to learn about the role before choosing between hands-on practitioner and manager paths.

Staff designers have already reached the level this book is talking about. Establishing a foundation at that level is a confusing and lonely process. Many staff designers are figuring it out on their own. This book will help you along the process of setting up excellent working practices so you can consistently meet or exceed expectations.

Senior designers often hit a wall in the process of seeking a promotion. They believe they are ready for the next level, but their manager has a different opinion. If you are a senior designer, this book will provide you with tangible examples that will illuminate your next steps and explain what your manager means when they use certain leadership buzzwords.

Senior UX practitioners who are at a fork in their career need to decide whether to become a super-senior designer or move into a management role. This book clarifies the differences between a staff designer and a design manager. It will show you how you can be a leader and support designers on your team without actually having any direct reports. You will be able to make a more informed decision about the future of your career.

Since the book covers expectation-setting along with the implementation of leadership tactics, it also benefits design leads, managers, and directors in the following situations:

- Shifting back to hands-on design and understanding how leadership skills will translate.
- Updating an organization's career ladder to best support the staff designers they hire.

The shift from management back to hands-on design can be uncomfortable and confusing. A staff designer is still a senior leader, but they do not have direct influence. This book will help former managers learn how to do great work at the staff level and affect organizational change and company strategy without managerial authority.

If you are a senior people manager, this book will assist you in creating an organizational design that empowers super-senior designers to do great work. Building empathy for individuals who work at this level will help you become a better leader and advocate for those in your management chain. Investing in research about the experiences of those who report to you will help you build a high-functioning team.

While this book is mostly targeted at designers, other UX practitioners will also benefit from its contents. The transition from senior to staff mostly requires a more strategic use of foundational skills. If you want to show up as the best version of yourself at work, improve at how you communicate, and facilitate better working relationships, you will get a lot of value out of this book.

What's in This Book?

This book covers the core focus areas, responsibilities, and archetypes of a staff designer at different company sizes. You'll learn what a staff designer is and is not, along with how the role of a staff designer compares to other positions such as design manager and senior designer. These expectations will help you position yourself appropriately within your organization.

With expectations established, you can begin to set a proper foundation that will empower you to influence your organization at scale. You'll learn about time management, project prioritization, and how to set boundaries so your energy can be used on more high-impact work. After you've set a solid foundation, you'll learn about the different types of people you'll need to connect with and how to start using the information you glean from them to make an impact.

Once you begin to make a positive change to the business, you'll need to work across others to increase your impact. You'll learn ways to delegate work to your fellow designers, along with how to maintain experience quality when you're on the hook for the quality of their outputs. This book also covers methods to document and communicate your impact. For those who are more humble, you'll also learn to toot your own horn without looking or feeling like a jerk.

Finally, this book will explore the super-senior career beyond staff. You'll learn options for the future of your career, including climbing the ladder or switching roles. In addition, you'll learn ways to reinvigorate yourself if you hit a standstill in your lengthy design career.

What Comes with This Book?

This book's companion website (https://rosenfeldmedia.com/books/the-staff-designer/) contains the book's diagrams and other illustrations—all available under a Creative Commons license (when possible) for you to download and include in your own presentations.

FREQUENTLY ASKED QUESTIONS

I've been a senior designer for a while. Why am I not staff yet?

There are several possible reasons why you haven't gotten promoted yet. One possibility is that the career ladder is not super clear, meaning expectations of the role haven't been given deep enough thought. In that case, you'll need to work with your manager to go over the expectations of the staff role and discuss the ways in which it differs from senior—we go over this in Chapter 1. You also might not be thinking far enough into the future about your product area. Visioning is discussed at length in Chapter 5. Finally, it's also possible that you are doing everything right, but your manager is unaware of your impact. You need to improve at communicating your impact more broadly if this is the case; Chapter 8 will help you out.

I'm a [lead/principal/other fancy title] designer. Is that the same level as staff?

Possibly! It depends on the structure of your organization's design career ladder. The lead level usually sits right above senior and is quite similar to staff, although it often includes a people management component. The principal level may be equal to staff, or it might be the level above staff. All of this is covered in Chapter 1. Additionally, the scale of your company might affect the needs of your role and affect how you might position yourself at a different company. Chapter 2 is all about organizational design.

My manager keeps saying I need to build influence. What do they mean?

One possibility is that you are probably a great collaborator to your peers, but you aren't connecting with senior leadership enough. This will require a heavy investment in your collaborative relationships, which we cover in Chapter 4. It's also possible that your ideas aren't getting very far—perhaps you aren't pitching in a convincing enough way. You can begin to advocate for your ideas once you have audited your workplace connections, and we go over that in Chapter 6.

I'm in so many meetings. How am I supposed to find time to make such a large impact?!

You need to befriend your calendar and get in control of your workload. Chapter 3 has lots of time management and work prioritization techniques that will help you free up focus hours. Another issue might be that you need to delegate the more tactical work so you can focus on larger-scale projects. Chapter 7 gets into the concept of scaling yourself so you can claw back time for more strategic efforts.

What is a vision and how am I supposed to make one?

A vision illustrates how a concept will play out in the near or far future. Designers can create visions for small features, product areas, or entire products. For example, you can make a vision that shows how your team's feature might be used by other product areas. Or you can make a vision that shows how your team's product area might evolve in two to three years. Chapter 5 outlines the process of defining and presenting a vision.

CONTENTS

How to Use This Book	vi
Frequently Asked Questions	ix
Foreword	xviii
Introduction	xx

CHAPTER 1

What the Heck Is a Staff Designer? — 1

What a Staff Designer Is	3
Level-Setting	4
Expectations of a Staff Designer	8
What a Staff Designer Is *Not*	10
A Staff Designer Is Not a Senior Designer	14
Nor a Manager	16
Staff Designer Archetypes	20
Architect	21
Tastemaker	22
Visionary	23
Platformer	25
Debrief	26
Activity	27

CHAPTER 2

The Impact of Org Design — 29

Org Design Hinders or Enables Success	30
Small Companies (0–99 Employees)	33
Focus	33
Speed	34
Team Structure	34
Setting Up for Success	36

Medium Companies (100–499 Employees)	36
Focus	38
Speed	38
Team Structure	38
Setting Up for Success	40
Large Companies (500+ Employees)	41
Focus	41
Speed	41
Team Structure	42
Setting Up for Success	44
Debrief	46
Activity	47

CHAPTER 3

Wrangle Your Time and Capacity   49

Make Time for Visioning	51
Befriend the Calendar	51
Stop Context Switching	56
Audit Each Meeting	58
Color-Code Calendar Blocks	59
Capture Your Workload	60
Build Your Backlog	64
Plan Projects	65
Set Healthy Boundaries	66
Estimate Capacity	67
Triage Tasks	70
Say "No" More	72
Debrief	74
Activity	75

CHAPTER 4

Nurture Your Relationships — 77

Build Your Network — 78
    Contributors — 79
    Observers — 82
    Approvers — 84
    Supporters — 85

Document Your Findings — 86
    Dynamics — 87
    Opportunities — 92
    Sentiment — 93
    Symptoms — 97

Debrief — 97
Activity — 98

CHAPTER 5

Drive Product Vision — 99

Leaving a Seat at the Table — 101
    Renounced Power — 102
    Lack of Insight — 103
    Fidelity Mismatch — 103
    Perfectionism-Driven Silos — 104

Take Back the Wheel — 106
    Power: Bring Curiosity — 106
    Insight: Increase Confidence — 107
    Fidelity: Progressively Upscale — 109
    Silos: Design in Public — 110

The Value of a Vision	111
What Is a Product Vision?	111
When to Propose a Vision	112
How to Make a Vision	113
Debrief	118
Activity	119

CHAPTER 6
Build Influence Without Authority   121

How Influence Works	122
Layer 1: Direct Control	125
Layer 2: Influence	125
Layer 3: Beyond Influence	126
Pick Your Battles	126
Business Priorities	128
Context Overlaps	128
Ownership Areas	129
Gather Your Research	130
Organizational Insights	130
Qualitative Insights	131
Quantitative Insights	132
Build Your Case	132
Observation	133
Proposal	134
Outcome	135
Share Your Case	136
Presentation	136
Delivery	138

How to Handle Disappointment	139
Appeal	139
Move On	140
Debrief	141
Activity	141

CHAPTER 7

Scale Your Impact — 145

Know When to Delegate	146
Find Work to Delegate	147
Approachable	149
Moderate Impact	149
Low Conflict	150
Low Complexity	150
Choose How to Delegate	151
Internal Delegation	152
External Delegation	154
Set the Bar	156
Techniques for Leadership	157
Handle Challenges	162
Debrief	163
Activity	164

CHAPTER 8

Show Your Value — 167

Why Humility Fails — 170
Manage Your Presence — 172
- Demeanor — 173
- Succinctness — 175
- Candor — 176

Manage Up — 177
- Set Expectations — 178
- Share Regular Updates — 181
- Broadcast Impact — 184

Debrief — 188
Activity — 188

CHAPTER 9

Keep Your Career Fresh — 191

Change Companies — 193
- Drive Critical Projects — 193
- Navigate Ambiguity — 194
- Work Across Others — 195
- Create Usable Experiences — 196
- Get Results — 196

Go Beyond Staff — 197
Switch Roles — 199
- Content Design and Information Architecture — 201
- Design Operations — 202
- Design Management — 203
- Product Management — 205

User Research	206
UX Engineering	207
That's All, Folks!	208
Give Yourself Grace	209
Rest Often	209
Follow the Fun	211
Debrief	211
Index	**212**
Acknowledgments	**222**
About the Author	**226**

FOREWORD

I've spent the last decade growing as a product designer across multiple start-ups, Facebook, GitHub, and now at Notion. At every company, it can feel a bit like starting from scratch: different titles, different definitions of "senior," and vague or conflicting expectations for how to grow beyond that level.

I vividly remember hitting a point in my own career where the next "level up" looked a lot more like middle management and a lot less like doing the work that made me fall in love with design in the first place. I watched many friends and colleagues make the transition into management, thinking it was the only path forward, only to quickly fall out of love with design.

Out of my own frustration, I started staff.design, a long-form interview project for people to tell their own stories about the challenges and skills required to excel as an individual contributor in the design industry.

But those interviews were just interviews. They weren't a playbook. And each reader still had to synthesize their own point of view about where to go next.

That's why I'm excited about this book: Catt Small's *The Staff Designer* is a clear, honest, and deeply thoughtful guide to navigating the challenges of advancing as an individual contributor. It adds shape to the invisible work and influence required at the staff level and shares concrete tools you can use to grow your impact without sacrificing your craft. Not only has Catt drawn from her own personal experiences to guide designers on this path, but she has also recruited stories from a wide range of design practitioners (both managers and individual contributors) to add context and nuance that is often missing in discussions about the designer's path.

It's not all about the pixels!

The strength of *The Staff Designer* is that it doesn't romanticize the role. Instead, it breaks down the subtle (and often frustrating) realities of working in complex organizations with actionable tips for scaling vision, impact, and value. Working at this higher level of abstraction often catches designers by surprise: suddenly growth isn't directly tied to pixel output, and it's all too easy to get stuck.

Whether you're trying to break through the "senior ceiling," or want to solidify your presence at the staff level, this book will meet you where you are and help you push past.

—Brian Lovin
Creator of staff.design and Product Designer at Notion

INTRODUCTION

Back in 2018, I was a senior designer struggling to understand what I needed to do to grow to become a staff designer. I had been at the senior level for several years at that point and had launched several of the highest-impact projects. These projects unlocked massive growth for my employer at the time.

Additionally, I had a successful public speaking practice and a solid presence on the design internet. But no matter what I did, nothing seemed to convince my manager that I was ready for the next level. The career ladder wasn't much help either, as I matched every criterion listed but still couldn't figure out how to get a promotion.

Then, in early 2019, I began to look for more resources to understand how I could keep growing. My friend and coworker at the time, Jessica Harllee, became a precious mentor to me. She was one of the first few staff designers at Etsy. At the time, she was the only woman in the role, and our backgrounds were similar enough that I felt I might be able to emulate her path.

I was inspired to invest in some reading, thanks to my conversations with Jessica, so I began looking for books about designing beyond the senior level. At the time, no books covered the perspective of an individual contributor, so I invested in a copy of *The Making of a Manager*[1] and *Radical Candor*.[2] My hope was to get some tips that might translate to my role.

I struggled in this situation for another nine months and then decided to look for a new role elsewhere. Luckily, another company saw the impact of my work and recognized my skill set. That's how I landed my first role with the official staff title.

In 2020, I began that new job and quickly learned how different the role of a staff designer truly was. Luckily, I was mostly prepared, but there were still many moments of awkwardness and impostor syndrome. Getting comfortable with senior leaders took time and learning to own my presence as a leader was a process.

---

1 Julie Zhuo, *The Making of a Manager* (Portfolio, 2019).
2 Kim Scott, *Radical Candor* (St. Martin's Press, 2019).

Around that time, a nifty website called staff.design popped up. Brian Lovin was a staff designer at GitHub then, and he interviewed other staff designers about their work. Their thoughts were so helpful as I developed my own practice as a super-senior designer. But then the site went dormant after a year. I kept waiting for someone else to make more content, but no one did. So, I began to reflect on what others might learn from my own experience.

After reflecting on my career and speaking to other super-senior designers, I recognized that I encountered several common challenges along my path to growth:

- **Shifting goalposts:** The career framework was nonexistent or unclear—meaning anything can be your job.
- **Context switching:** Too many people ask you for support with different projects, leading to a fragmented calendar with little focus time.
- **Stonewalling:** Team leaders and executive stakeholders regularly ignore your input, even if your recommendations are correct.
- **Shadow impact:** Direct collaborators know you're amazing, but leadership doesn't understand what you do. You feel under-recognized.
- **Structural blockages:** The company's org design requires you to work in ways that are not conducive to the kind of impact you were hired to make.

And that's why we're here! This book is the culmination of not just my time as a staff designer, but my 15 years of design work in general. Each chapter is designed to build on the prior one so that you can strengthen your knowledge. You'll get to amplify your power through practice activities that have been created with the explicit intent of building your skills as a staff-level leader.

Core principles of this book are relationships, sustainability, and scale. Each of these principles is critical for performing successfully and consistently as a leader. Staff designers may have no direct reports, but they must adhere to these principles as much as any other leader would.

Relationships will help you succeed in your role. Designers are facilitators; we take inputs, analyze them, and create solutions based on what we learn. We don't usually build those solutions—this requires a team. If you're going to get the best solutions built, you'll need to get others to buy into your vision.

Sustainability is crucial to consider because the best products are built in a sustainable fashion. They focus on the long-term goal. Designers can help teams look past the dopamine rush of low-hanging fruit and get them to see the full forest. By asking the right questions, designers can bring human connection back into strategic conversations. They can also contribute to a learning mindset that results in a more resilient product development team.

Scale matters because it is the goal of most companies that hire designers. Whether they employ a small or large number of individuals, most capitalist enterprises intend to keep growing. A staff designer will help their company scale in size, product offering, user count, or any other form of impact that is relevant to the business. As time passes in their role, they will also have opportunities to scale their impact from direct, hands-on work to collaboration across many people. Operating with scale in mind will help you prepare for your future steps.

You'll notice that these principles influence content in every chapter of the book. Each chapter includes extensive research and insights from not just me, but also over 20 product design experts at the staff and above level. This is the book for learning how to lead and succeed as a staff designer. Whether you're currently staff and struggling to get your bearings or are senior and unsure what you must do to keep growing, this book is for you.

Throughout the book, you'll also get to read exclusive interviews with designers who are already at the staff level. I've also included a few perspectives from principal designers and senior managers, so you have a well-rounded set of insights about both your closest partner and the potential for even further growth beyond staff.

You're well on your way to becoming a successful individual contributor. Remember to be open, curious, and patient with yourself. Growth takes time, and you're a whole human—not just a designer. Onward!

CHAPTER 1

What the Heck Is a Staff Designer?

What a Staff Designer Is	3
What a Staff Designer Is *Not*	10
Staff Designer Archetypes	20
Debrief	26
Activity	27

Staff designers can be incredible assets to a business. They have the power and experience to tackle exceptionally complicated user experience problems that unlock major revenue opportunities. And employers are noticing! In August of 2025, I searched for "staff product designer" roles on LinkedIn in the United States. More than 100 relevant roles were posted within seven days alone. Organizations are hungry to hire talent at this level, offering immense sums of money and incredible perks to the right individuals with relevant experience.

Despite its ubiquitousness in today's job market, the staff designer role is still relatively new. The World Wide Web may be over 30 years old, but the career track for super-senior designers of internet-connected devices is a much more recent creation. Until recently, most designers were expected to shift into management at a later stage in their career. Design management has a stable foundation and decades of history built on various pre-existing forms of leadership theory.

Once a designer becomes senior, the management track is made available to them. They can become a middle manager, eventually reach the director level, and potentially become a VP of design or chief design officer with enough effort. The progression is straightforward and predictable. Many designers do not want to manage people, though, and we should not force them to. Unhappy managers come with cascading negative effects: they create harm at scale by passing their pain along to every individual in their management chain.

The software engineering career track has a model for success that the design industry must follow. A considerable number of engineers have refrained from transitioning into people management, which led to the definition of the staff engineering career path. Engineers can stay close to code while tackling more and more ambiguous architecture problems as they advance in level. There are dozens of visible super-senior engineers who regularly share their experiences. The design field has to catch up.

In the first year of teaching my staff designer course, I heard over 150 variations of the same predicament: the staff designer career path is opaque, and employers are unsure how to create ideal conditions for people who take on this role. The role of a staff designer is currently defined on a company-by-company basis. Some organizations don't even have this path available at all, leaving individuals to advocate for themselves when they hit a career ceiling. Staff designers are set up to fail when the expectations of the role are unclear to everyone, including the companies who hired them.

When career ladders across companies vary to this degree, designers lack the clarity to excel because they are cannot calibrate appropriately. And this difficulty is amplified when a designer attempts to transition their skill sets between companies—the standards at one employer may be completely different from another. In the open-source world, companies share technologies with each other because it pays off in faster advancements for all. The same applies to the staff designer role: universal expectations and standards can be shared between organizations for the benefit of the entire industry. These standards will help companies of all sizes establish better design practices, resulting in so many great things: more efficient internal operations, faster onboarding and seamless career growth for designers, and ultimately better outcomes for users.

This career path is still relatively nascent. We are progressing, and that progress comes with growing pains. Growing pains are frustrating and uncomfortable. On the bright side, it also means you're part of the conversation. The decisions you make as you chart your own path will help create clarity for future designers.

This book is a stake in the ground to set expectations for the role of a staff designer, and we're starting with this chapter. You'll find that most staff designers have strengths in some of the expected skills more than others. The reality is that there isn't enough time in the day for designers to excel at everything all the time, and it's unreasonable to be equally brilliant at every facet of the role. You'll learn enough detail to decide where to invest your energy. You'll also learn what a staff designer is *not* so you can set appropriate boundaries with collaborators and stakeholders. Let's battle career ambiguity together!

What a Staff Designer Is

A staff designer *is* a super-senior designer who does *not* manage individuals. Nonmanagers are colloquially referred to as *individual contributors* (ICs). Staff designers are individual contributors, but they are likely to be responsible for the design strategy and user experience quality of a major product area. Despite having no direct reports, staff designers are often extremely influential and valuable members of senior leadership.

FIGURE 1.1
A common design career ladder based on insights from 75 organizations such as Netflix, Pinterest, and Shopify.

Level-Setting

I've reviewed career ladders from 75 tech companies, including major organizations such as Airbnb, Datadog, Instacart, Lattice, Microsoft, and Shopify. I merged these insights with my own personal observations from past employers and synthesized them into the most common career ladder you'll find (Figure 1.1). I will mostly be concentrating on the role of a product designer, because this is currently the most common title that companies assign to digital experience designers. Product design, as defined here, is the combination of information architecture, visual design, and strategic product thinking. Much of this information is applicable to other related practices such as UX design, content design, and service design.

> **TIP LEVEL-MATCHING ACROSS SUPER-SENIOR TITLES**
>
> Please note that a designer may be evaluated distinctly at different organizations, depending on the context of each company. If the level right above senior is *principal* in one organization's career ladder, that may be considered the same as *staff* at another organization. Scale also affects this translation; for example, a principal designer at a 100-person company might be a senior designer at a 1,500-person company due to the scope and complexity of working at a larger organization. However, if a designer makes a lateral move to a company of a similar size but each company uses unique titles for the level above senior, the designer would likely be assigned the same level.

Typically, the design career ladder starts at entry-level (*design intern* or *apprentice*). It then moves to *junior designer*, then mid-level (just *designer*), *senior designer*, followed by *staff designer*. *Staff* is the most common title at the level above *senior*—two thirds of the companies I reviewed used it as a title. The remaining few either used the term *lead* (Figure 1.2) or *principal* (Figure 1.3). Nomenclature is often determined by the preferences of leaders at a particular organization.

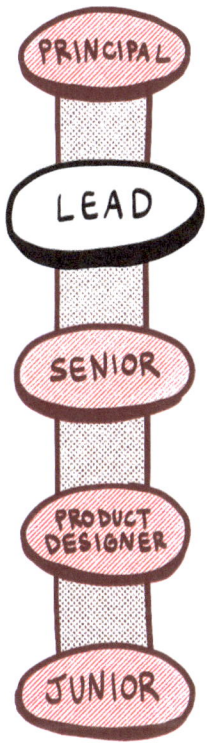

FIGURE 1.2
The career path for a company with lead designers.

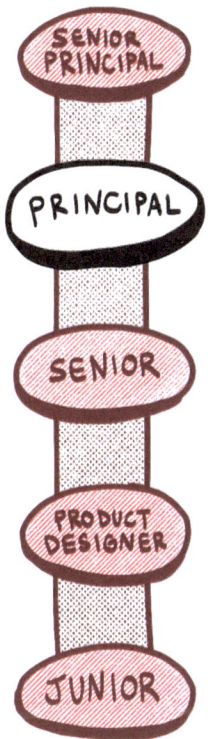

FIGURE 1.3
If the level above senior is called *principal*, it is equal to the staff level.

Lead designers function similarly to *staff designers* in many ways. Lead designers are usually hands-on designers. Similar to staff designers, they usually own a major product area. However, they often do what is called *player-coach* work, which means they do both tactical design activities (player) and manage at least one other designer (coach). They are expected to deliver high-quality, hands-on product design work *and* function as a low-level people manager. In comparison to lead designers, most staff designers work at a similar altitude without the challenges of managing others.

The *principal* title is often used above the staff level, but it may be interchangeable with staff at smaller companies with less hierarchy. Beyond the staff level, a designer can grow to become either senior staff or principal (Figure 1.4). Different companies use these terms interchangeably. And if a company is mature with 1,000 or more employees, they may have senior principal, architect, and distinguished designer titles (Figure 1.5).

FIGURE 1.4
The career ladder at a company where the principal is the role above staff.

FIGURE 1.5
While rare, some organizations do offer at least one level above principal.

Very few designers reach levels above staff. However, the skills required are an extension of those needed to jump from senior to staff. The shift from senior to staff is the largest hurdle a designer faces in their career, due to the increase in ambiguity and political maneuvering. Therefore, I will be concentrating on the staff designer level for most of this book. If you're curious to learn about the level above staff in more detail, we'll cover that in Chapter 9, "Keep Your Career Fresh."

IN THE REAL WORLD

YOKO SAKAO OHAMA

Yoko Sakao Ohama is a Staff Product Designer at a major eCommerce marketplace. Yoko and I met during our overlap at this company, where she has worked for over eight years. Throughout her 14-year career as a professional designer, Yoko has only been a hands-on designer. She joined the company as a mid-level designer in 2016 and then was promoted to the senior level in 2017. After five years, Yoko was promoted to the staff designer level.

"I was a senior designer for a long time—I felt like I was good where I was for a while. Then I got put onto projects that were increasingly complex, and I enjoyed that. I started seeing gaps that existed in the product that would need someone at the staff level to observe, articulate, and fix."

Many companies require individuals to perform at the capacity of a higher level for six to nine months before they can receive a promotion. This was the case for Yoko. "They let you get practice before you officially get put into the role. By the time I got promoted, I was ready to be promoted."

Yoko also believes that high-ranking roles like staff and principal designers require a business case. "Those roles need to be identified and advocated for. In order to create the job description that hires for that role, the company needed to be in a good enough spot where they knew my product area was expanding. They needed someone who could operate at a staff level, and I grew into the role."

Expectations of a Staff Designer

A staff designer is an individual contributor who has moved beyond the senior title and level. They are essentially super-seasoned designers who can wrangle complex, ambiguous projects without much hands-on support. The expectations of designers at this level are very high, and they are often brought in to put out major fires. One might even call them modern-day unicorns!

Similar to lower-level designers, staff designers are usually responsible for the end-to-end outcome of at least one large project. This is still a hands-on role, so staff designers are expected to create design artifacts just like other designers. A staff designer will create whatever is necessary to drive the most impact: wireframes, diagrams, mock-ups, and other forms of design documentation are all fair game. Because staff designers are so experienced, stakeholders expect them to create higher-quality artifacts than lower-level individuals on the team—in less time. In fact, staff designers are expected to have such excellent skills that they can help other designers level up, too!

What makes a staff designer unique is that they also provide direction for others in many forms. At the beginning of a project, a staff designer might create a visionary prototype that gets broken down into chunks. Depending on an organization's size, the staff designer will either shift to executing on a chunk of work or hand parts of the work to other teams so they can scope and build the work. If other teams take on the work, the staff designer is usually expected to review and provide feedback on the progress to ensure that it is executed as intended.

Staff designers are excellent verbal and written communicators who are deeply connected to the hands-on design practice. They use design to predict potential futures, support strategic decision-making, and create great outcomes for their company's customers. Key focus areas for designers at this level are systems, craft, and business strategy (Figure 1.6).

Each of these focus areas must be balanced with the other. Lots of designers fall into using visual design as a crutch by making showy mock-ups without considering how things connect. Other designers miss major opportunities for long-term revenue wins by overly concentrating on short-term gains. Context is important, so make sure to invest equally in these focus areas as you grow.

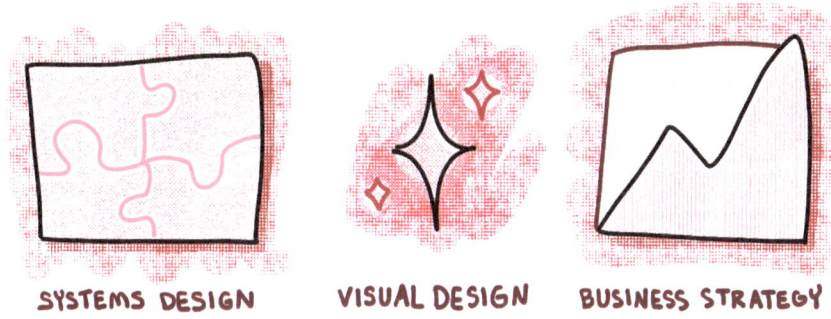

FIGURE 1.6
Staff designers must excel at systems design, visual design, and business strategy.

Staff designers are clearly expected to do a lot! That's because they have a long track record of high-impact work and many years of real-world experience. Don't feel rushed to get to the title if you're not there yet.

On the other hand, if you are leveled at senior but feel like this whole chapter describes you already, you might be ready for the next level. You'll likely need to make a case for a promotion. You can figure out ways to advocate for yourself in Chapters 6, "Build Influence Without Authority," and 8, "Show Your Value."

Systems

Systems thinking is the ability to create connections between ideas and concepts in a system using artifacts such as user flow diagrams, affinity maps, and wireframes. It helps designers cut through ambiguity to show teams how everything connects. Staff designers own elaborate projects that tie to key business initiatives. Systems thinking is key to understanding customers at scale—and finding new business opportunities. Customers will give you a lot of feedback. How do you and your team decide what to work on? As a staff designer, you must influence these decisions, and you will be a better influence when you can see how everything connects.

Craft

Craft, also known as *visual and interaction design skills*, is crucial for all product designers. Staff product designers are expected to both deliver and model fantastic craft. Therefore, staff product designers must be better at visual design than senior designers. These super-senior individuals push the bar upward, and they raise the floor for design quality across the team. A solid foundation in color theory, typography, and other graphic design principles is required to succeed at this level. The best staff designers know how and when to use visual design as a tool; they see when to follow the grid and when to break it. They use the right fidelity designs to motivate leaders, supercharge teams, and inspire customers.

> **TIP CRAFT IS SUBJECTIVE**
>
> Different companies have varying opinions regarding what excellent craft looks like. One organization's bar for visual design might be higher than another's. Each company prioritizes particular qualities of visual design over others. Your visual design skills will be evaluated uniquely at every place you work. If you want to work at a place with products that are considered to have top-tier visual designs, expect the bar for craft to be much higher.

Business Strategy

Staff designers also have a magnificent sense for business strategy, or the ability to define and execute on a plan to achieve certain outcomes on behalf of the organization. This is what makes a designer at this level a worthwhile expense—they prioritize their energy to ensure that they are always working toward an outcome that benefits the business by satisfying its customers. It's the opposite of designing for design's sake: every decision you make is optimized for impact. You'll learn more about prioritization in Chapter 3, "Wrangle Your Time and Capacity," and influencing business strategy in Chapters 5, "Drive Product Vision" and 6.

What a Staff Designer Is *Not*

Companies think a staff designer is many things. There are some very real overlaps with other roles. But a staff designer is definitely *not* a manager, nor are they a senior designer. These super-senior individual contributors own (rather than contribute to) major initiatives, and they are not responsible for any direct reports!

A DAY IN THE LIFE OF A STAFF DESIGNER

Staff designers have a lot on their plate. Each day, week, month, and quarter looks very different. I'll attempt to flatten my years of work in the role into one discrete example—a pretty standard Tuesday. Since this is an example, it's possible (and likely!) that your lived experience will be different and unique.

10 a.m.: Reconnaissance

I open my computer and look at my calendar to assess what the day will look like. Then I check my task manager and see a reminder to spend time developing a pitch for a new project idea later in the day. Finally, I open my company's chat platform and see a group message sent by the VP of product: They have an idea for an urgent new project that impacts the area I own.

I put on my architect hat and look at my capacity. It's possible that I'll have availability for this project, but the goals are unclear. I invite the VP, my PM, and my engineering lead to a quick, impromptu kickoff.

10:30 a.m.: Kickoff

During the meeting, I duplicate a handy template for a kickoff document. I use it to define expectations of the work, facilitating the conversation between my teammates and the VP of product. By the end of the half-hour discussion, it's clear that much more thought is required before the project can begin.

11 a.m.: Planning

I write up a proposal for a several-day brainstorm that concludes with a research session. I believe this will answer the questions that came up during the kickoff and give the team enough information to move forward with confidence. After an hour of preparation, the proposal is in splendid shape. I send the proposal to the meeting's participants for review.

12 p.m.: Lunch

Since staff designers are senior leaders and have influence over the behavior of others, I am strict about modeling healthy calendar hygiene. I take an hour-long lunch every day. Half the time is spent devouring a proper meal. Then I go for a walk to clear my mind. All that thinking is a lot of work!

continues

A DAY IN THE LIFE OF A STAFF DESIGNER (continued)

1 p.m.: Critique

The team has design critique twice per week, and the first one is on Tuesday. This area has several projects in flux at the moment. Every designer's contribution must add up to a first-class, end-to-end experience. I give clear and thoughtful directional feedback, celebrating wins while also offering opportunities for improvement.

2 p.m.: Pair design

Right after critique, I have weekly 30-minute 1:1s with individual designers on the team. While they aren't my direct reports, I am accountable for the quality of their design work. These are deep dives into work discussed during critique. We explore options together, and I leave each collaborator with just enough feedback to get them unstuck. At the end of each session, I provide a little time to answer general questions. One designer needs more time at the end of their session, so we schedule additional time for tomorrow afternoon.

3 p.m.: Inbox zero

I open the company's chat program for the second time today—I check it three times per day in total. I see that people tagged me in several of their area's channels. After 30 minutes of answering questions, everyone is unblocked. I close the chat program and open my music player.

3:30 p.m.: Focus time

Now that the major meetings are out of the way, I spend time exploring potential concepts for a pitch using the company's preferred design tool. I'm considering some ideas that might be worth investing in, as I noticed a few issues that need a more holistic solution. Right now, the team is resolving these issues in a piecemeal fashion. A better solution would require my area to collaborate with another one.

5 p.m.: Influence

The new ideas are quite interesting! Now I need to craft a pitch that will get others on board. During the final 30 minutes of my available focus time, I record a video walkthrough of the proposed concepts and share it with my PM for feedback.

5:30 p.m.: Ramp down

With the day nearly over, I check to see if there are any comments on the brainstorm proposal I wrote earlier. There are. After resolving the comments, I schedule the brainstorm for next week. Next, I reopen my task manager and update my to-do list. Finally, I open the chat program for the last time and write an update for my teammates.

6 p.m.: Log off

Once again modeling healthy calendar hygiene, I say good night to the team and log off.

A Staff Designer Is Not a Senior Designer

Senior designers have less responsibility than staff designers. The scales of their projects are very dissimilar. Table 1.1 outlines the key differences between senior and staff designers. See the Catt's Corner sidebar for specific examples based on real-life projects.

TABLE 1.1 DISPLAYING DYNAMIC CONTENT

Activity	Senior Designer	Staff Designer
Influence strategy	Short-term	Long-term
Large scopes of work	Execute	Define and execute
Partners with	ICs and managers	ICs, directors, executives
Design foundations	Strong	Excellent
Team craft	Meet the bar	Uplevel the bar
Mentor other designers	Junior and mid-level	Junior, mid-, and senior-level
Work without direction	Most of the time	All of the time

While both senior and staff designers can be trusted to be relatively autonomous, they are disparate. Staff designers are expected to influence long-term strategy (12+ months), whereas senior designers usually focus on immediate priorities. Staff designers often collaborate with executive leaders, directors, and senior ICs. Senior designers, on the other hand, usually work with other junior and senior peers.

A staff designer might be responsible for a project that has a critical impact on the organization, meaning they work at a higher altitude. Meanwhile, a senior designer might contribute to a slice of the staff designer's project—driving a manageable portion of the impact considering their level of experience.

CATT'S CORNER
Workloads and Altitudes

Students in my course often ask for examples of projects that show the differences between the staff and senior levels. Staff designers often work on the definition and execution of strategy. Senior designers usually focus on execution, often contributing to a project defined by others above them (say, a staff designer or people manager).

Here are five examples of projects that I worked on as either a senior or staff designer:

- An entertainment company wanted to create a new subscription product. They tasked a *staff designer* with helping them to determine the scope of the customer experience. Once the full scope was determined, I and several other *senior designers* owned slices of work that contributed to the full scope. The staff designer ensured that our work met the company's quality bar.
- An eCommerce marketplace wanted to redesign the analytics that platform sellers use to determine how buyers find them. They asked me as a *staff designer* to determine the new architecture and vision for the redesign; then they assigned several *senior designers* to the project as support at key points in the process. I assigned ownership of specific parts of the analytics experience to those senior designers as we worked toward launch.
- A task management software company wanted to build out a new product area for goal management. They hired me, as a *staff designer*, to define the ideal scope and MVP for the product. Once we scoped out the MVP, several *senior designers* on other teams were assigned portions of the MVP scope that overlapped with areas their teams owned. I connected with those designers on a weekly basis to ensure that their work contributed to a cohesive experience.
- A venture corporation hired my employer (an agency) to help them combine two products they acquired into one new experience that was better than the sum of its parts. The agency assigned me as a *staff designer* to the project to scope out the architecture of the new experience and then assigned *senior design support* to help me execute on portions of the scoped work.
- A B2B software company wanted to redesign their admin console, which was outdated and intricate with over 80 pages of settings. I, as a *staff designer*, mapped out a new information architecture, created an updated visual system for the product area, and prototyped what the future experience would look like once all the changes were made. Several *senior designers* took sections of the prototype and created engineering-ready designs for all the additional user states that were necessary to launch the redesign.

As you can see, *staff designers* often receive ambiguous and complex scopes of work, create clarity by illustrating the future, and then shift to execute on their strategic direction and ensure the quality of those who contribute to the final experience. Meanwhile, most *senior designers* are assigned a moderately complicated scope of work and receive direction from above. A senior designer can often show that they are ready for the next level by improving at navigating more ambiguous, difficult scopes of work and building the ability to create direction for others. ■

Staff designers are also expected to have a deeper knowledge of design foundations at an industry-wide level. They have a splendid track record of delivering high-quality work and a broad toolkit of exercises and usability insights from past experience that they can pull from to move projects forward. Senior designers have a terrific grasp of these foundations, but they are expected to *meet*, not *uplevel*, the design quality bar.

Both senior and staff designers are expected to mentor others. But senior designers mentor junior and mid-level designers. Staff designers have enough experience to mentor senior designers on the team. They might even mentor design managers, directors, and people in other positions, such as content design. You'll learn more details about mentorship and facilitating exceptional design in Chapter 7, "Scale Your Impact."

Besides scale and scope of work, the staff designer title also comes with much higher expectations in comparison to a senior one. Much of this comes down to experience that comes with time. Don't rush to fit into this role! Concentrate on expanding and stabilizing your aptitude with each project you take on. The change in scope and responsibilities will come as your confidence increases.

Nor a Manager

Staff designers are leaders. Design managers are leaders. Although they are both leaders, *staff designers are not design managers*.

A manager's job is to build a talented team and ensure that its members output great work. A staff designer's job is to create positive business outcomes through design with a combination of hands-on work and direction. The major overlaps that exist between the two roles are mentorship, craft, and strategy (Figure 1.7).

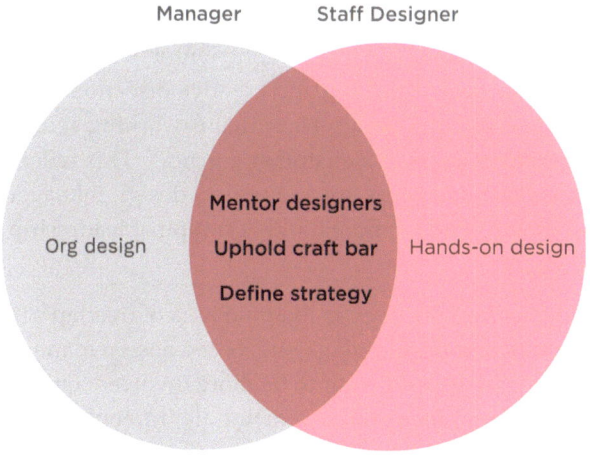

FIGURE 1.7
Managers and staff designers partner to mentor designers, ensure the design quality bar is met, and define strategy.

Staff designers do not have direct reports. They should, however, mentor designers to guarantee that the quality of design work is as high as possible. With a staff designer on the team, design managers can focus on professional development and team output. The staff designer can do pair work with less senior designers on the team—akin to old-school apprenticeships. This process elevates the quality of the team's output.

A staff designer and design manager are both held accountable for the quality of the team's work. They both must be present during craft-centric team rituals, such as design critiques. Managers often have a high-level view across multiple projects and focus on sharing key strategic details that help to unblock their direct reports. They ensure that the right work is shown at the right time, and all teammates have opportunities to give feedback. Meanwhile, a staff designer usually owns the details for a specific major project and delivers the kind of nitty-gritty feedback that might otherwise be categorized as micromanagement.

Design leaders often reflect on the differences in these roles as they refine their career ladders. In March 2025, a design leader named Ben Martin wrote about this process in a post on his blog, *Designing the Gap*. He noted that "Staff designers focus on mentorship and high-level execution, while design managers drive quality through team development and career growth." In 2020, Peter Merholz, writer of

Org Design for Design Orgs, also observed that more companies are hiring people into "super-senior individual contributor" roles to support leadership with "strategic thinking and creative direction that can make sense of the effort of a design team that is working across many products, or distinct parts of a customer journey." This reflects my lived experience: Staff designers are partners to design managers and directors, and they provide a more hands-on type of leadership that is rooted in the craft.

As Peter alluded to, business strategy is the final area of overlap between staff designers and design managers. Some design managers care deeply about strategy and want to be more involved. Others want to be less involved and focus on supporting their people. The latter is rarer in 2025, so staff designers are expected to share ownership of the strategy and partner appropriately.

The industry has ebbed and flowed in terms of its preference for people-focused management. Most companies currently prefer managers who can balance their focus on people with a heavy investment in strategic and creative direction. Cap Watkins, VP of Product Design at Lattice, reflected on his own experience with this shift in a blog post called "The Rebalancing of Design Management:" "If managers aren't actively driving business outcomes using all their knowledge and expertise, … then all the people-focused work we all care about doing isn't going to matter because the product won't exist for long."

Managers are commonly held accountable for strategic influence and impact. If you're on a team with a manager who has lots of strategic opinions, both parties equally own the strategy. It's important to hash out roles and responsibilities with your team's design manager so the boundaries are clear. For example, perhaps the manager owns the project's execution plan and timeline while you drive the actual definition of the project's strategy.

Other than these overlaps, these roles are complementary. Managers concentrate on org design and team cohesion. Staff designers execute design and make sure the end-to-end experiences they own are cohesive. Whenever you find yourself in a situation where the lines are blurred, get out a pen and draw some boundaries with your design manager.

IN THE REAL WORLD

ANGIRA SHIRAHATTI

Angira Shirahatti is a Product Design Manager at Datadog. I met Angira in 2017 after she made the transition from product design intern to mid-level product designer at Etsy. She reached the senior level by 2020 and then transitioned into a graduate program to explore the world of academia.

For Angira, graduate school felt like an extreme version of her potential future as an individual contributor. "I didn't feel the fulfillment that I was expecting." Knowing the parts of design work that engaged her most—persuading stakeholders, setting strategy, and creating an environment where others can thrive—she became curious about the world of people management.

After grad school, Angira re-entered the workforce with another senior role at Gusto, and then eventually landed at Datadog. She took on larger and larger projects, but the feeling of existential dread stuck around. A year into her tenure, she officially made the switch and became a product design manager. While the transition required a substantial change to her self-image as a builder, she's enjoyed the change overall. "Helping others grow and develop in a safe and fun environment and showing up for people when things are hard has been incredibly rewarding."

The shift from individual contributor to people manager required Angira to taper off nitty-gritty design work and focus more on mentorship and coordination. Her experience as a designer has come in handy for helping her reports navigate different situations. "A big part of my job now is being able to predict how a project will go, mitigating risks, and giving advice, so my personal experience as an IC is something I pull from every day."

Angira is still deeply involved in design strategy; however, she's had to shift her focus from leading one project to many. "Since each of the designers on my team owns a whole product, I can't possibly know all the details of all the things they're working on, and it's not productive for me to try." Instead, she ensures that her team is set up for success with a well-balanced portfolio of work. Her day-to-day work centers on the facilitation of collaboration between designers, and she is happiest when everybody on her team has a project that's both fulfilling and helping them progress toward their career goals.

Staff Designer Archetypes

Every staff designer is unique and brings distinct value to an organization. However, there are four archetypes that most designers fall into. These archetypes are architect, tastemaker, visionary, and platformer (Figure 1.8). The first two archetypes may be familiar to you, as they are core competencies of a designer. The latter two are unique to designers at the staff level and above, as they help individuals scale their impact.

FIGURE 1.8
Each archetype combines the core competencies of a staff designer in varying ways.

Most designers have a dominant archetype that they fit into most of the time. It's natural to have an affinity for certain archetypes over others. For example, I am an architect by default! I love to learn how users work. Cracking a meaty problem gives me joy, and I love the feeling of meeting or even exceeding customer needs.

Throughout my career, my architect instincts have helped me resolve bigger and bigger challenges. However, I can still put on the tastemaker hat when necessary. And I've vastly improved at the skills required to meet expectations of both the visionary and platformer archetypes.

Great staff designers can flex into the other archetypes as necessary, wearing them like hats. While some designers are innately adept at certain skills, most skills are learned and polished over time. If you find yourself naturally gravitating toward one or more of the archetypes after reading the descriptions that follow, consider investing energy into strengthening skills that fit the ones you are least comfortable with.

Architect

Do you know those folks who can take any mess and turn it into order? The ones who can look at a user flow, list out all the objects it contains, and categorize them by type? That's an architect! Architects are excellent systems thinkers who solve complicated experience issues. Like those who design buildings, architects create the structure and scaffolding for well-constructed user experiences.

Strengths

Companies value architects because they ensure that teams are solving the right problems and identifying the right solutions. Architects combine systems thinking with business strategy to ensure that an experience is cohesive. They use information architecture to help customers perform critical tasks, ultimately helping the team reach business goals.

An architect might create user flow diagrams that align teammates on the number of screens to include in an end-to-end experience. Or they may put together a wireframe that shows how content will be grouped in an experience. They use the arrangement of ideas, objects, and concepts to reflect business and user priorities.

Shadows

While architects are great for making sense when everything is in disarray, people who default to this archetype are in danger of becoming overly logical. There is an inherent nonsensical quality to the aesthetic part of our role as designers, and architect-dominant designers are the first to get frustrated by visual craft conversations. If you're an architect, you should continue investing in your inner tastemaker, so you stay connected to user interface design as a skill.

Build This Skill

Architects have a solid foundation in user experience (UX) design and human-computer interaction (HCI). Designers who want to improve their architect skill set can invest in fundamentals such as information architecture, systems diagramming, and user research. Designers can build these skills by attending workshops and events through organizations such as UXPA, IxDA, Interaction Design Foundation, and Nielsen Norman Group.

Tastemaker

There's always one designer on the team who has unreal and exceptional visual design skills. That person is your quintessential tastemaker: a designer with an eye for interface, style, layout, and typography. They experiment and play with interfaces, sending messages through the visual treatment of certain objects on the screen.

Strengths

Tastemakers combine visual design with business strategy to influence user behavior and guide customers through an experience with the intention of meeting certain outcomes. They help translate a company's brand into experiences that customers want to use. Other designers who witness the work done by a tastemaker may feel inspired to improve their own craft, leading to a collective increase in the quality of experiences the company produces.

Similar to the way that first impressions matter, the visual layer of a product is immediately apparent. Companies often prioritize tastemakers because their work draws people in. Design savants and novices alike can easily appreciate the outcome of a tastemaker's meticulous nature, who needs every pixel to be in the right place.

Shadows

Organizations love a masterful tastemaker. But designers must expand their focus, so they don't get pigeonholed. People who "make it pretty" are not as valuable if they can't ensure that their solutions solve the right problems. If you are a natural tastemaker, consider deepening your practice as an architect and visionary.

Build This Skill

Contrary to popular belief at companies, not all designers are natural tastemakers. While some will have an affinity for visual design, many individuals will need to invest extra time and energy to excel at it. I personally have a professional background in graphic design, but this archetype requires the most significant effort for me to maintain. Visual design is a continuous puzzle for me to complete.

Despite anything folks have told you in the past, it's possible to keep building the muscle for craft. Tastemakers have a good foundation in visual design and interaction design. Designers who want to improve their tastemaker skill set can invest in typography, color theory, grid layouts, and user experience prototyping. Many traditional universities offer continuing education for tastemakers. Designers such as Elizabeth Lin and Matt D. Smith (MDS) also offer their own courses on this subject.

> **TIP CRAFT FOR OTHER ROLES**
>
> In one cohort of my staff designer course, a researcher asked me how to translate craft to other roles. I believe most subsets of the user experience practice have their own definition of craft. For example, a UX researcher's craft might be defined as the way they design and conduct their research. A content designer's craft might be measured by how well they construct information within an experience. If you're not sure how "craft" translates to your role, consider aligning on the definition of the term with your design team.

Visionary

Think of a designer you know who gets immediate support for their solutions. They tell such a compelling story that the whole team—even senior leaders—rally behind the idea. That designer fits into the visionary archetype.

Strengths

Visionaries are designers who use storytelling for influence. They are especially powerful for cases where executives need to buy into an idea. They combine systems thinking, craft, and business strategy to show how a solution will impact customers (Figure 1.9). Visionaries create artifacts that build the kind of confidence necessary to enact major product changes.

Less senior designers often struggle to illustrate how their work ladders up to plans for the future. This work is not about creating an artifact. It's about using a deep understanding of the customer and business problems to expand on potential futures.

Without a doubt, the visionary archetype is one that most differentiates a staff designer from a senior one. By delving into alternate worlds, staff designers help leaders build confidence in major investments. When managers speak of influence, the visionary skillset is the one they're most often referring to. We'll cover this more in Chapters 5 and 6.

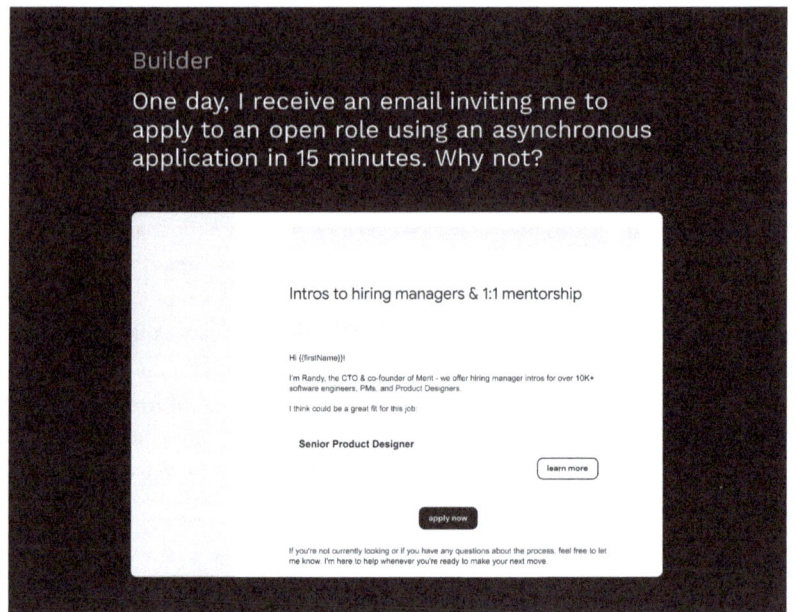

FIGURE 1.9
Visionaries illustrate potential futures for the team to help them make and communicate strategic decisions.

Shadows

Natural visionaries must watch out for excessive vision work. Every project doesn't deserve an elaborate presentation; sometimes, the work just needs to get shipped. If you have a proclivity for creating grand scenarios of possible futures every time you begin a new project, consider the context. I reserve vision work for times when the team lacks direction.

Build This Skill

Visionaries have a foundation in the same skills as tastemakers and architects. However, they have also invested lots of time into their storytelling, presentation design, and public speaking abilities. While some institutions do offer training in public speaking and storytelling, many designers can build these skills through observation of others and consistent practice.

Platformer

The platformer creates repeatable documentation and components that other designers can use. They combine systems thinking with craft to generate user experience frameworks and guidelines (Figure 1.10). This is key to being able to scale yourself as a staff designer (see Chapter 7). I also gravitate toward this archetype because documentation helps me ideate and iterate faster.

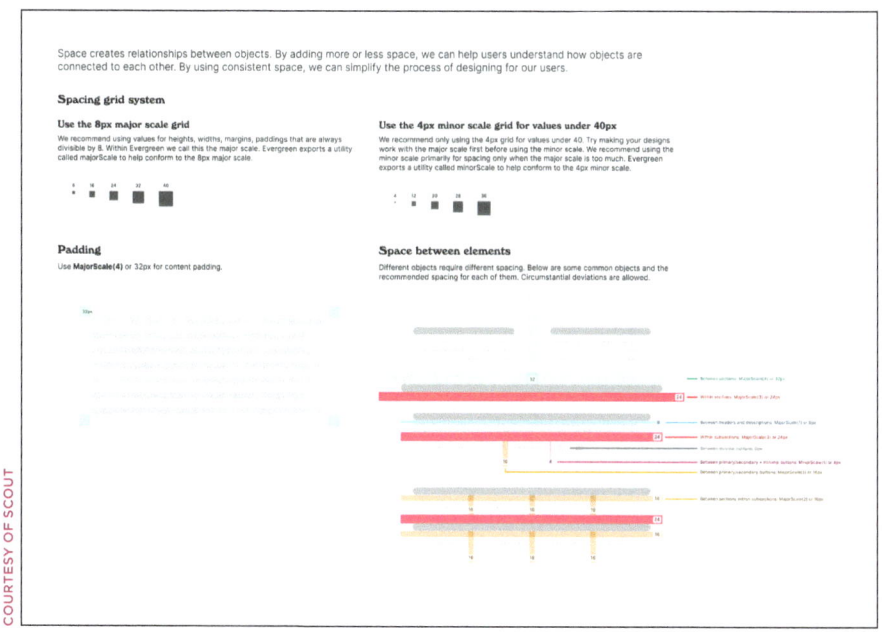

FIGURE 1.10
Platformers create artifacts that help others contribute to large-scale projects in a cohesive manner.

Strengths

If you have an affinity for design systems but want to continue working on a traditional product development team, you might be a natural platformer. Platformers are fierce advocates and partners to design systems. They often connect easily with design systems designers and continuously contribute to the system, making them laterally influential.

Shadows

Platformers enjoy creating structure, just like architects. However, too much documentation and process can be distracting—and can ultimately lead to being pigeonholed as a design systems designer. Get other members of the team involved in contributing to and maintaining the documentation you create. Co-creation builds a sense of shared ownership.

Build This Skill

Designers who fit the platformer archetype have a foundation in the same skills as platformers and architects. Those who want to improve their platformer skillset can also invest in learning to create design systems documentation. Designers can build these skills by directly contributing to their company's design system, taking classes by design systems thought leaders like Dan Mall, and attending events such as Clarity, a design systems conference.

Debrief

Staff designers are hands-on senior leaders who influence the future of the company in a way that is distinct from both senior designers and design managers. They use a combination of systems thinking, visual design, and business strategy to inspire and mobilize teams at scale. Each staff designer has unique strengths, but most fit into a dominant archetype as their default operating style. Staff designers must regularly invest in skills that enable them to change archetypes based on context. By changing archetypes as necessary, staff designers adapt to meet the needs of ambiguous projects and make a major impact.

Activity

Identify three times where you fit into each staff designer archetype. Then reflect on which type resonates with you the most. Use the results to determine your growth areas.

Archetype	Example 1	Example 2	Example 3
Architect *Systems designer who simplifies complex flows.*			
Tastemaker *Creates high-quality designs that uplevel the bar.*			
Visionary *Storyteller who aligns teams around a vision and leads them to execute.*			
Platformer *Creates concepts and patterns other designers and teams can use.*			

Which archetype do you identify with the most?

Which archetype are you the least like?

Which archetype does your company need most often?

CHAPTER 2

The Impact of Org Design

Org Design Hinders or Enables Success	30
Small Companies (0–99 Employees)	33
Medium Companies (100–499 Employees)	36
Large Companies (500+ Employees)	41
Debrief	46
Activity	47

Organizational (org) design is the intentional arrangement of a company's teams, groups, and reporting lines to ensure overall success through the output of its people. Every company has an org design, whether intentionally curated or plotted under duress. Founders usually set the initial design of a company when it is established. Some follow best practices that were created by org design researchers and professionals with relevant experience. Others chart their own paths, pushing the envelope with new structures; these experimental designs may have positive or negative results.

As a business scales, its founders hire managers. Those managers then modify the org design to meet the needs of the business, hiring more people and creating new teams as necessary. Depending on how intentional the founders were, the initial design of a company may have ripple effects that last for years beyond its founding. For example, Stripe, a financial services company founded in 2010, continued its anti-hierarchy philosophy by refusing to use official job titles as it scaled to over 4,000 employees in 2023.

In a Harvard Business School course, "Leading Change and Organizational Renewal," Professors Michael Tushman and Charles O'Reilly highlighted four fundamental elements of org design: organization-level processes, department structure, manager-report ratio, and hierarchy. Managers aim to optimize each of these elements to improve operations and scale. These design changes, referred to as *re-orgs*, are often made in sweeping changes that can cause disruption and growing pains.

Individual contributors do not usually have power over their company's org design because they are not responsible for human resourcing. That means you won't be directly participating in the practice of organizational design as a staff designer, but it will affect *how* you work in many ways. Designers who are aware of org design's effects deliver more effective design solutions that are more likely to reach their intended audience.

Org Design Hinders or Enables Success

The number of organizational layers you navigate on a day-to-day basis will change the way you communicate with your manager and senior leadership. As shown in Figure 2.1, if you're three degrees from the Head or VP of Design at a 1,000-person organization, you'll have to go through three levels of reviews at every step required to launch a major project. That amount of effort could add several extra weeks to your project, and a re-org could take months or even years to implement!

Companies of different sizes also hire staff designers for different reasons, so you must bring the right perspective to the role based on the scale of the organization. As a staff designer, you are expected to assess your situation and position yourself appropriately to best succeed. If you don't adapt, you'll be viewed as a high-cost yet low-impact employee—this makes you less valuable and puts you at risk of being laid off.

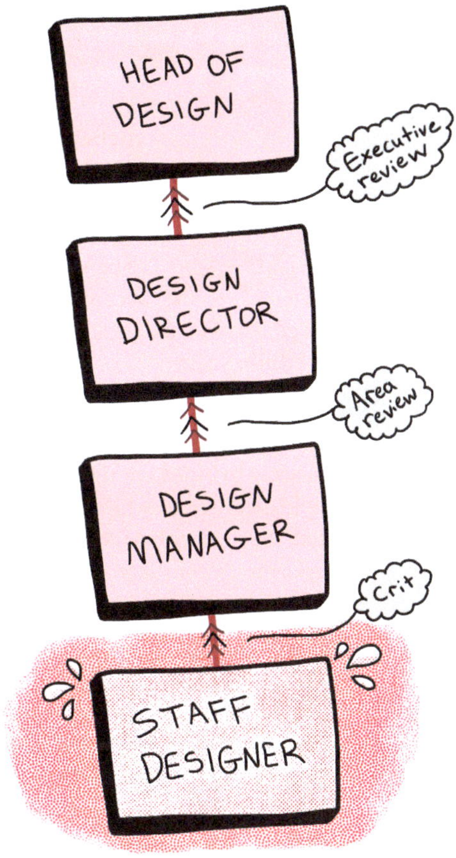

Every organization is unique, but for the sake of brevity, I will focus on companies at three key sizes:

- Small (0–99 employees)
- Medium (100–499 employees)
- Large (500+ employees)

These are generally defined as the three major stages in an organization's growth life cycle. If you work on software for businesses, these scales will likely be very familiar to you. The size of a company determines the number of management layers it contains, which therefore impacts the role of staff designers.

FIGURE 2.1
Reporting to the wrong person can slow down progress on your work.

The number of layers of management you have to go through determines the speed at which you can deliver high-impact work. In an ideal world, the number of management layers you regularly communicate with would inversely relate to the amount of impact you are expected to make. At a small company, there will be fewer people and therefore fewer managers; conversely, a large company will have more employees and therefore require more people to lead them. If you report to the right leader, you can spend more time doing what really matters and focus less on busywork.

IN THE REAL WORLD

MICHELLE KWON

Michelle Kwon is a Staff Product Designer at Flowcode. I met Michelle nearly 20 years ago when we both attended a graphic design undergraduate program at the School of Visual Arts. We both went into UX and ultimately became product designers. Michelle has since worked across various start-ups, helping them to go from the small to medium scale.

Michelle usually joins companies when they are small and quickly builds relationships with her collaborators. "One of my main focuses is to build enough rapport to establish myself as someone people can turn to when they are facing challenges. I want people to trust me enough to come to me for design debates, chitchat, and support with conflicts. It's personal and also professional."

Michelle reports to the Senior Director of UX Design, and there are two other designers on the team. Her working relationship with her manager is heavily collaborative: "I cater to what he's missing in the landscape of what he's trying to achieve. He's a very visionary person, and when he hears a concept, his mind goes to a million different places all at once. How do we move from one million different places to something that's more tangible? What is the next step?" By filling in information gaps and creating momentum, Michelle provides clear value to her manager and helps the entire team deliver better work to their customers.

Since Flowcode is a young company, Michelle positioned herself as an expert leader who could add valuable structure to the team's design process. "I am very consistent with communicating design principles to junior and senior designers. I provide clear and consistent guidelines and feedback that help the team home in on the key metrics and value props that we want to deliver." While they only have a handful of designers on the team, Michelle's design guidance ensures that they all move in the same direction.

> **START-UPS COME WITH GROWING PAINS**
>
> According to the United States Small Business Administration, 99.7% of companies with paid employees in 2023 had fewer than 500 people on the payroll. Most employers are small or medium businesses. Many of these companies are figuring out their structure as they grow. Some employers may reference common knowledge and proactively prepare the design org to scale, but others will only create as much structure as is necessary at any given moment.
>
> At least once in your career, you will probably work at a company with a less-than-ideal org design. When you encounter poor org design, you will be expected to navigate communication challenges with grace. Designers who report further down the chain will need to deliver high-impact work while aligning leaders across multiple layers of management. Meanwhile, designers who report directly to executives will have limited support, so they'll need to self-manage more often. Both situations can be helped using techniques covered in Chapter 8, "Show Your Value."

Small Companies (0–99 Employees)

Companies with 0–99 employees are small. These are what people usually think of when the word *start-up* is used. Most organizations out there fit into this category, and most of them will not succeed. Like a roller coaster, the start-up journey can be fun if you have the stomach for it!

Focus

Small companies usually only have a team of 1–5 designers. Nicole Dominguez, a design engineer with over 15 years of experience at small and medium companies, noted that "designers end up with multiple hats" in organizations with less than 100 people. They might design an end-to-end flow for a new product area one day and then make new icons or create a new slideshow template the next day. Nothing is above a designer's pay grade at a company of this size, and everyone is expected to pitch in to fill the gaps.

While the scope of work is large, the staff designer title is rare because there is so little hierarchy on the team. Companies of this size will often hire their first designer as a "head of design" or "founding designer." This person will either be expected to grow into a management role or eventually hire the person who will manage them.

Designers at small companies must focus on driving impact as quickly as possible. This is often done by generating experiences that meet—or ideally exceed—customer expectations. This output will result in the foundational growth the company needs to break even and ultimately succeed.

Speed

Due to their minute scale, small companies are usually very intense and fast-paced. Many companies of this size are mission-driven, meaning their purpose is widely communicated internally and deeply connected to daily operations. The CEO is likely heavily involved with day-to-day decision-making. Scope is often large, and career growth is quick.

Team Structure

Because the company is small, the design team is microscopic (Figure 2.2). Sometimes designers at small companies report directly to the company's founders since there is no need for a head of design. Companies with closer to 99 people might have designers report to a head of design, head of product, or CTO. There will likely be one product team that handles many tasks across the entire product, and everyone will be expected to stretch to fit the needs of the company.

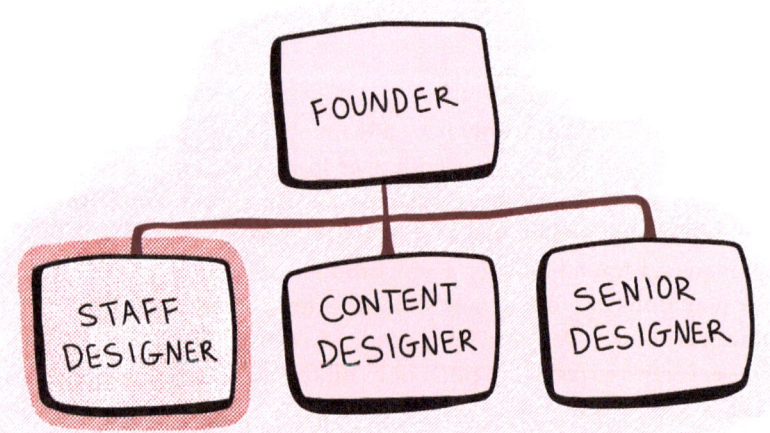

FIGURE 2.2
An example of a small company's design team.

Collaboration with Executives

Designers at start-ups often collaborate directly with the founders. Even if a designer does not report to a founder, they will likely communicate with each other on a regular basis since founders are so heavily involved in daily operations at small companies. Managing upward is a common challenge that designers at small companies face, as the founders usually have strong opinions about the direction of the product. Building a collaborative relationship and working in public can help founders understand and respect the design process.

Design Leadership

Many small organizations do not have any design managers. This is both a blessing and a curse, especially for staff designers. A staff designer is self-sufficient and therefore does not need much support from a design manager. However, they will need to handle many interpersonal situations without the cover of a manager. Designers at start-ups must be prepared for the emotional labor required to maintain healthy working relationships.

Cross-Functional Collaboration

Small start-ups are lean operations. There may be a head of product, one or two product managers, an engineering manager or CTO, and a handful of engineers. User researchers and other valuable roles are usually considered a luxury that the company cannot afford to hire.

Like design at companies of any size, designers at small companies are usually expected to communicate constantly with their product and engineering partners. Due to the small scale of the company, the span of ownership will be large. Designers will be expected to support their team's needs to make quick decisions, and pivots might occur at any time. Product and engineering partners can benefit from an opinionated design partner who will help them visualize the implications of their actions. By getting involved with strategy, designers at start-ups can become leaders instead of being typecast as paintbrushes.

Setting Up for Success

According to a 2024 report from the United States Bureau of Labor Statistics, over 20% of small businesses fail within the first year. Most start-ups believe in removing dead weight quickly. Within the first few weeks of employment, a designer at a small company must make an impact and work with relative autonomy. Taking on a self-sufficient mindset is key to survival and success.

When a designer onboards to a small company, their priority will be to build a connection with the founders and builders. The team will be on the smaller side, and there likely won't be much documentation. The design career ladder will also likely be nonexistent or immature, so expectation-setting must be done with each individual on the team. Stakeholder interviews are crucial for success at an organization of this size, as high-quality working relationships are often the only way to gain the historical context necessary to get up to speed. We'll cover more details about building great relationships in Chapter 4, "Nurture Your Relationships."

Designers at small companies must continuously balance quality and speed. Organizations of this size usually have limited runway, so teams are working against the clock to turn a profit or meet their next funding milestone. While designers should always set and reinforce boundaries, lots of flexibility and comfort with compromise will be required for success. Sarrah Figueroa, a staff designer with experience at multiple start-ups with eventual exits, recommends clarifying your scope of expertise and offering to hire contractors for projects outside that scope. This approach helped her ensure that she consistently operated within the realm of her expertise while learning new skills through observation.

Medium Companies (100–499 Employees)

While companies at this size are still classified as start-ups, they are much more established. A company at this size is either a cult classic or in the process of becoming well-known. It's that company you've been hearing about a lot lately, like Perplexity, Writer, and Replit in 2025.

MANAGING YOURSELF

Small companies have little to no hierarchy. The upside of this setup is lots of autonomy. The downside is that you will be expected to own your career direction. Your manager will not have time to invest deeply in your career growth. They may not even be a designer!

If you join a small company, do not expect your manager to actively support your career growth. You will need to create your own path to success. Reflect on career direction with your cross-functional peers to see how they refine their skills. Lean on external resources such as podcasts and blogs. Build a network of peers outside your company by attending local and online events. Aggregating the knowledge of others can help you determine your own next steps.

Here are some resources you might find useful when you have little managerial support. Some of them are free, and some of them cost money. Before spending money on self-development, check with your manager to confirm whether or not your company provides discretionary funds for this purpose.

Podcasts
- *Design Details* by Brian Lovin and Marshall Bock
- *Design Life* by Charli Prangley and Femke van Schoonhoven
- *Dive Club* by Ridd
- *Per My Last Email* by Sara Wachter-Boettcher and Jen Dionisio
- *Technically Speaking* by Harrison Wheeler

Blogs
- *Ask a Manager* by Alison Green
- *Leading Product Design* by Leslie Yang
- *Proof of Concept* by David Hoang
- *The Beautiful Mess* by John Cutler
- *UX Design Weekly* by Kenny Chen

Events
- AIGA Design Conference
- ConveyUX
- SmashingConf
- Tech Week
- UXDX

Focus

Since the organization's goal is growth, a staff designer at a medium organization needs to focus on scale. The company wants to grow enough to move from moderate success to tech unicorn status. They've taken a huge bet by hiring such an experienced and expensive designer. The designer will likely be handed the reins to a major product area or initiative.

Speed

Like small companies, medium companies move quite quickly. However, they usually have more structure because there are enough people employed to require additional processes. It's an ideal situation—teams move relatively fast, and there are a number of resources due to the company's growth.

At a medium company, the CEO is usually working to build enough financial and customer growth to enable an exit. An exit could be going public through an IPO, getting acquired, or any other situation that would result in a major payout for those who have been involved since the early days. This is a make-or-break stage where the few breakout start-ups that continue to exist either fizzle (most likely) or reach hockey stick growth that causes them to become unicorns (rare).

Team Structure

Staff designers at medium companies often report directly to the company's head of design (Figure 2.3). This closeness creates opportunity for staff designers to move with agility and build toward the hypergrowth the company requires for its exit. They will likely work across a small number of other designers in addition to driving their own projects.

Collaboration with Executives

At a medium company, the CEO will likely be less involved in day-to-day operations. They may opt to focus instead on external efforts, such as brand building and investor relations. If they have properly hired a powerful team, they will only need to step back into direction work at critical moments in the company's growth journey.

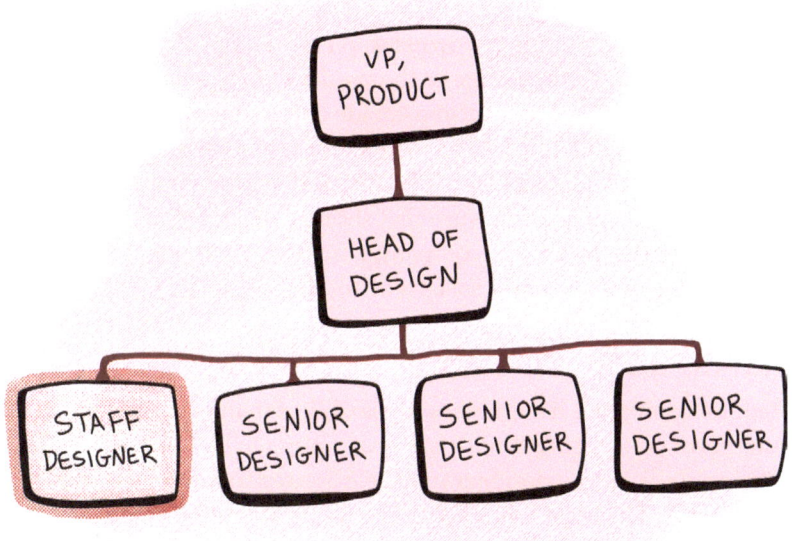

FIGURE 2.3
An example of a medium company's design team.

Staff designers may still occasionally interface with the CEO for important strategic projects. Annual planning is one such example of a project that might require CEO input. Otherwise, staff designers will probably not partner with the CEO on a day-to-day basis, as other major priorities will take precedence at an organization of this scale.

Design Leadership

Organizations with more than 100 people will have enough designers to employ at least one design manager, who will likely be referred to as the *Head of Design*. Staff designers act as secondary leaders who support the Head of Design with maintaining the design team's culture. With their many years of experience, staff designers are often pillars of the design team. In addition to delivering hands-on work, staff designers at medium companies mentor others, facilitate workshops, and drive visioning exercises that propel big ideas forward.

Cross-Functional Collaboration

The span of a staff designer's ownership is still large at a medium-sized company. While the core of the product likely won't change, product and engineering decisions made at this stage will have far-reaching effects that impact the product for years. By illustrating the potential long-term outcomes of product and engineering decisions, designers can help teams ensure that short-term success doesn't hinder progress toward the company's mission. We'll talk more about driving a vision in Chapter 5, "Drive Product Vision."

Setting Up for Success

Medium companies have a bit more flexibility than small companies in terms of onboarding time. Instead of aiming to be functional within a week, a new designer may be expected to ramp up over a few months. The team should have documentation that empowers the designer to get up to speed and ship their first project by the end of their first quarter.

A medium company has more resources, but it's still a start-up. The team has to move with agility to ensure that the company survives. During the first 90 days in a new role, a staff designer must build solid connections with their head of design along with product, engineering, and research counterparts. The design team is likely comprised of a handful of individuals, so they must also connect with each of their design peers. A consistent flow of information is critical at this stage, as major decisions could be the difference between an eventual IPO or bankruptcy.

> **TIP ONE SIZE DOESN'T FIT ALL**
>
> Most businesses may be small, but they are all unique. Every company has a distinct set of people and problems. Some designers apply a one-size-fits-all approach to their work, insisting a particular process must be followed, regardless of the context. Meghan Logan, Staff Product Designer at Thrivent, recommends against this approach. "If you're on a small and lean team, the odds of your org giving you weeks to run research are slim to none, and the odds of your org having a budget to run research or implement analytics is, maybe, even lower."
>
> Staff designers adapt to support the needs of the businesses they work with. They recognize the constraints that affect their situation and act accordingly, using relevant techniques from

their robust toolkit to help the team progress. For example, if your employer doesn't have the capacity for a full-scale research study, you might lean on feedback from the sales and support teams or schedule time with a few of your company's most engaged customers.

At small and medium companies, there will be more situations in which you must trust your gut and take calculated risks. Meghan noted that "You'll need to lean heavily into design heuristics and best practices that give you the data points you'll be missing otherwise." When qualitative insights are scarce, research firms like Baymard and Nielsen Norman Group can help designers at start-ups fill gaps with confidence.

Large Companies (500+ Employees)

Established companies, many of which are household names like Google or Netflix, are usually large. This is a huge umbrella of organizations that can contain anything from high-headcount series D, pre-IPO start-ups to major corporations with tens of thousands of employees. Although this range is gigantic, most staff designers at large companies operate similarly. For example, my staff designer position at Asana (1,000 employees) was similar to that of Lil Chen, another staff designer, when she was employed at Google (10,000+ employees). The only difference was the number of layers we had to navigate; she had to present to over twice as many senior leaders because there were many more levels of management.

Focus

Steady momentum and progress are the name of the game at a large company. Staff designers at large companies must create alignment. They work across product managers, engineers, other designers, researchers, and a plethora of stakeholders to achieve major outcomes. Companies can only achieve steady growth by continuously moving the needle, and staff designers are expected to make that happen.

Speed

Large companies operate more slowly and methodically than both small and medium companies because they have more employees—and much more at stake. Design teams at large companies are usually extensive with dozens or even hundreds of designers. Staff designers at large companies will have many resources to pull from, but also lots of stakeholders to manage.

CEOs of large organizations are usually removed from the day-to-day happenings, instead opting to direct their many employees by setting a clear vision. The company is either about to go public or already public, so their main concern is investor relations. Operations at the company may be tied to quarterly earnings reports, which must show consistent or increasing growth. The company requires and rewards predictable and steady growth with calculated risk-taking when necessary.

Team Structure

At a large company, staff designers usually report to directors or senior managers (Figure 2.4). They are responsible for either a major product area or work on a horizontal team that looks across the organization. This is the scale at which most org design challenges tend to occur.

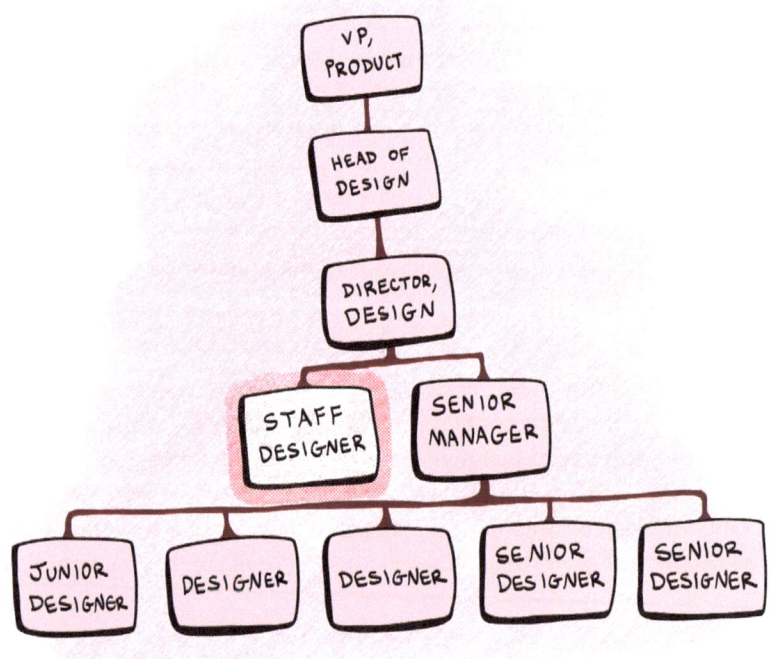

FIGURE 2.4
An example of a large company's design team.

Collaboration with Executives

While some CEOs care deeply about design, the CEO of a large company is usually extremely busy. Most staff designers will only engage with their CEO a handful of times per year, if at all. At this organizational scale, much of the work can happen independently of the CEO because the team should be full of astute senior leaders. The CEO's time is precious, so they will likely only be consulted before launching major initiatives that block the company's progress.

Design Leadership

Staff designers at large companies may find themselves further away from senior leadership than is necessary for the optimal pace of execution. In this situation, they will be slowed down by the additional reviews and meetings that are required to move through their many layers of management. Removing even one layer of management can substantially increase the designer's velocity since they will have one less presentation to schedule and prepare for.

Leaders are usually reluctant to make disruptive modifications to a company's reporting structure, especially if it means taking on more direct reports. Managing people is a lot of work at any level, and senior leaders would prefer not to add more to their busy workload. Therefore, time and patience are large components of corrective org design advocacy. Change is almost never immediate; in fact, it might take months or even years to see a correction. Org design modifications are often only made when the pain is acute enough to consistently and heavily impact a team's velocity, therefore slowing down the company's growth.

> **TIP** **NAVIGATING ORG DESIGN CHALLENGES**
>
> As a staff designer, the only way to navigate poor org design is to communicate proactively about the situation. By building relationships with senior leadership, a staff designer can identify the person to whom they need to report for the smoothest delivery of work. Since they are an individual contributor and not a manager, they are not responsible for the organization's design.
>
> Through direct collaboration and consistent feedback, staff designers can give their senior leaders the insight and awareness to make appropriate changes. Jason Huff, Director of Product Design at a major content publishing platform, recommends being both persistent and patient when influencing the shape of the

team and your position in it, including who you report to. "Don't simply influence your manager directly. To succeed in a large org, you need to be influential to leaders above your manager." The person you want to report to must believe that you have the skills to lead at a higher level, so you must show them that evidence while gracefully maintaining your relationship with your current manager.

Cross-Functional Collaboration

Some staff designers at large companies work directly with a single product team responsible for a major product area. In this case, they will be paired with a senior product manager and engineering team lead or manager. The scope of work at a large company might feel more minute than that of a small start-up; however, the implications are likely much more complex. Any change the team makes will likely have far-reaching effects that impact millions of users and have serious financial repercussions. Therefore, a staff designer is expected to be a strategic partner, helping their product and engineering teammates make the most effective decisions to reduce user-facing complexity and encourage consistent, predictable growth of key metrics.

Other staff designers at large companies operate within an agency model. In this situation, they are temporarily paired with different teams or groups to facilitate the most impact possible. Each project will be different, which means there is a high amount of variety. Staff designers who work on horizontal teams must be comfortable with unpredictability and thrive in ambiguity, as every quarter will look very different. They may even have to seek out their own new projects, continuously justifying the rationale for their role to exist. High levels of self-direction and conflict management are required to excel in this kind of position.

Setting Up for Success

Large companies usually do not expect to see results from hiring a new staff designer within their first three months. Rather, they intend for the designer to completely acclimate to the environment and gather enough historical context to bring new ideas to the team. New staff designers must focus their energy on building the foundation for collaborative relationships and identifying opportunities to make the most impact.

Due to the scale of products at large companies, staff designers usually require much more time in their tenure to affect major change. By the third month, the designer may start work on a high-priority project, but they likely will not ship anything substantial for at least six months. Designers often remain at companies of this size for two or more years due to the amount of time it takes to make a major impact.

> **TIP** SCALE REDUCES SPEED
>
> Staff designers at large companies must have a muscle for patience, as change and influence will take much more time. Edwin Morris, Staff Designer at Datadog, joined the company as a Senior Designer. Over the span of two years, he was promoted into the staff role and built an excellent track record of shipping effective work that gained him credibility with his team.
>
> People tend to stay at large companies for longer amounts of time. This means your connections multiply in value as the years go on. As Edwin grew, his coworkers did, too. "You might meet someone as a PM who, four years later, is a director and can help you get headcount for a design-led project." After five years of relationship-building and consistent delivery of high-quality work, he was able to use his influence with VPs and executives to get engineering support for an initiative he proposed.

CATT'S CORNER
Move Quickly and Thoughtfully

Large companies naturally move at a slower pace than small and medium companies due to the amount of coordination required across business units and between management layers. This difference in pacing can give product teams more time to explore and validate potential directions. The flexibility can be both a blessing and a curse, as the lack of urgency can lead individuals to lean into perfectionism and become more lax with time management.

My advice: don't let the sludge get to you. I've worked at companies of all sizes, and momentum is equally important everywhere. The more efficiently you operate, the more people look at you as someone who gets the work done.

Perfectionism can be useful. Meticulously honing a design can lead to an increase in experience quality. But perfectionism is often harmful, preventing teams from actually delivering value to users. I fight perfectionism with the power of scrappy start-up energy.

You can bring scrappiness to a large company in the right contexts. For example, I get creative with how my team and I gather insights. Can we reference existing research to make certain design decisions? How can we leverage internal users along with insights from the sales and customer support teams to move forward? Having more resources doesn't mean we should use them when we don't need to.

Want to inject some speed into a team at a large company? I recommend coming to the conversation from a place of curiosity. People are more likely to hear you when you are looking for opportunities rather than being critical. ■

Debrief

A staff designer's experience and responsibility will vary by their organization's size and design. The more complex the organization, the more management layers they will have to navigate. If a designer is at an awkward place in their org's design with either too many or too little management layers, their success will be much harder to recognize. The best way to improve the situation is to set clear expectations with your collaborators (covered in Chapter 3, "Wrangle Your Time and Capacity"), build great relationships (covered in Chapter 4), and share feedback with the goal of influencing change (covered in Chapter 5). Eventually, when the pain is severe enough, management will adjust your reporting chain so that agility can increase.

Activity

Illustrate your reporting line in your current or most recent company's design org. Look at where you're placed and reflect on how it impacts your work. What do you notice?

What's working well for you in this org structure?

What issues might be caused by this org structure?

If you are in the proper place within your company's org structure, congratulations! Not quite? Answer the following questions.

Considering your organization's hierarchy, who is the ideal person for you to report to?

What are 1-3 steps you can take to improve your situation?

If you don't have regular meetings with your ideal manager, schedule a 30-minute 1:1 with them as soon as possible. We'll discuss ways to build relationships in Chapter 4 and influence organizational change in Chapter 6, "Build Influence Without Authority."

CHAPTER 3

Wrangle Your Time and Capacity

Make Time for Visioning	51
Capture Your Workload	60
Set Healthy Boundaries	66
Debrief	74
Activity	75

Like all human beings, staff designers have a limited number of hours in the day. Yet they deliver a magnitude of value and impact equal to that of multiple designers. This is only made possible through heavy levels of self-awareness, regulation, and boundary-setting with others across the axes of time and workload capacity. As shown in Figure 3.1, it's not possible to quickly deliver visually appealing, super-usable designs without burning out. Therefore, staff designers must work with their teams to prioritize and identify the most important work to be completed within a certain timeframe.

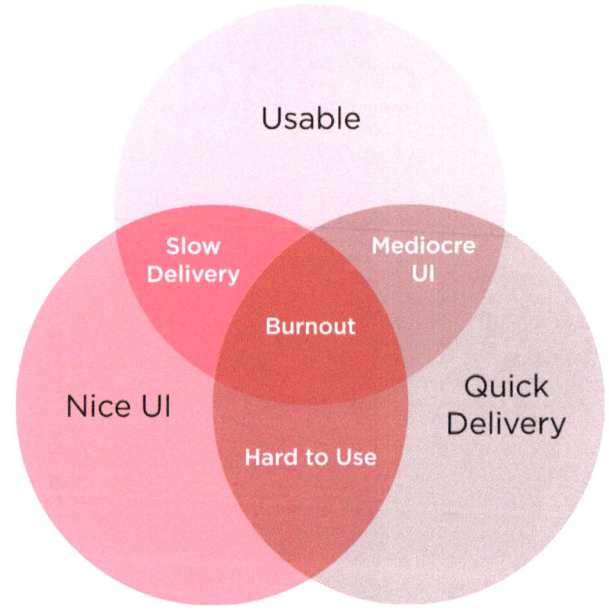

FIGURE 3.1
Designers must set clear expectations and help teams decide when to prioritize speed or experience quality.

Time determines the number of hours you have to complete your workload. For example, let's say there's a staff designer who works 40 hours per week. They have 20 hours' worth of meetings on average, and they take one hour to eat lunch every day. That gives them 15 hours to complete their workload. The designer must optimize those 15 hours to ensure that most of them can be spent on focus work, and they must claw back additional hours when possible by declining to attend irrelevant meetings.

A *workload* is the number of tasks a designer must complete during a given amount of time. Using the previous example, if a designer only has 15 hours each week of time to focus on completing their workload, the number of tasks in that workload must be adjusted appropriately. The designer's team must account for the amount of time the designer will have when considering the amount of work they can take on as a whole. They must also prioritize tasks in that workload to agree on which tasks deserve the most time.

Staff designers constantly pull the levers of time and workload capacity to deliver the most impact possible. At any given moment, they will adjust either the amount of time they can spend on a task or the amount of work they plan to deliver within a period of time. By communicating their availability through these two lenses, they help their teammates understand the effort that goes into creating elegant designs.

> **TIP YOUR MILEAGE MAY VARY**
>
> If you're a time management wizard like me, some of the advice in this chapter may seem obvious. But you'd be surprised at the number of people who don't follow these tips. I've presented this information to over 120 senior and staff designers in my staff designer course, and it's proven to be consistently insightful. Even if some of it sounds obvious, I still recommend you go through this chapter and think deeply about how you manage your time.

Make Time for Visioning

Time is a critical resource in the world of business. Designers are therefore expected to be cognizant of every working moment. Four ways to make more time for vision work include befriending the calendar, reducing context switching, auditing meetings, and creating a color system for calendar invites. Together, these methods help designers optimize their time and have a better overview of their work week.

Befriend the Calendar

Designers are visual people. Staff designers visualize how they use their time because it helps them see trends and proactively plan their time. Sunnie Sang, a high-ranking designer at a large-scale social media company, takes time at the beginning of the week to organize her calendar and plan how she will spend her week. This approach ensures that she has an overall sense of what her week will look like and gives her the chance to block off focus time.

Sunnie does this because collaborators will usually book any amount of open time available, assuming it is free because it is available. Blocking off focus time protects it from being covered by lower-priority meetings. Many designers create recurring events for focus time each day so they can generally guarantee they have enough time to create high-quality work.

> **IN THE REAL WORLD**
>
> **LIL CHEN**
>
> **Lil Chen** is a Staff Product Designer with over 15 years of digital design experience. Lil and I met in 2012 when she was a Web Designer at TED. She transitioned fully into product design at Google, working her way up to a lead level over seven years. She joined Discord in 2023.
>
> During her time at Google, Lil learned a lot about managing her time. "Being firm with drafting and sticking to milestones helps me make sure I can keep engineers unblocked." She works with the team to agree on approximate deadlines for design-related tasks in any given project. "It's about understanding the key parts and how to sequence the work. If we decide on the approach by a certain time, I can spend the rest of the time doing refinement. While engineers build the larger infrastructure, I can work on pixels in peace."
>
> Lil also learned a lot at Google about setting appropriate expectations with peers when she foresees roadblocks. "I found success in overcommunicating the implications of design changes and how they impact a project's timeline." However, Discord is more flexible about timelines, so she needed to change her communication strategy. "They focus on the best product we can possibly ship. So here, it's about talking with my team, telling them where I am in the process, telling my manager what concerns I notice preemptively." This ensures that the team can course-correct and give Lil the space to refine her design work.
>
> Lil's change in communication style still gives people the time to react accordingly, prioritize appropriately, and make early changes that can have positive downstream effects. But she adapted to the needs of the business, focusing less on shipping by a predefined due date and more on crafting high-quality user experiences. The shift between how Lil approached expectation-setting at both Google and Discord illustrates the flexibility that is often required when a designer switches companies—a technique that works well in one organization might not fit the needs of another.

TIME MANAGEMENT TOOLS

Designers often ask me to recommend the best calendaring software out there. The reality is that everyone's context is unique, but I do have some suggestions. Here are some examples of tools that designers can use to manage their time better. Choose one of each to unblock the hours necessary to create work at the staff designer level.

Scheduling

Scheduling tools set the foundation for your time management toolkit. These calendar tools help designers get a sense of their day. Designers can use these tools to schedule recurring focus time and manually assign colors to calendar invites.

Options include:
- Fantastical (Figure 3.2)
- Google Calendar
- Notion Calendar
- Outlook Calendar

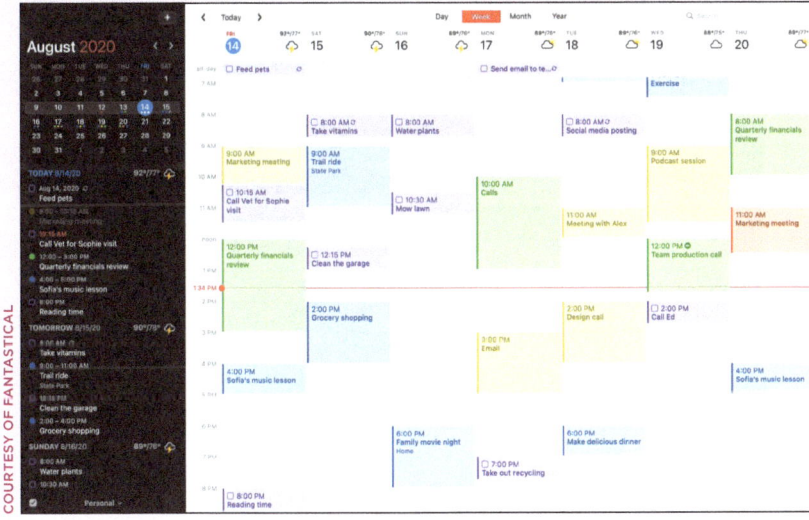

FIGURE 3.2
Designers who use scheduling tools like Fantastical to manage their time are more effective.

continues

TIME MANAGEMENT TOOLS (continued)

Automation

Automation tools empower designers to optimize their time without manually making updates. These tools can automatically assign colors to calendar invites and schedule meetings based on preset rules. Many of them can also schedule focus blocks and reschedule meetings to optimize time without anyone having to lift a finger. They usually also come with an appointment calendar that enables individuals to communicate their availability with external clients.

Options include:
- Clockwise
- Motion (Figure 3.3)
- Reclaim

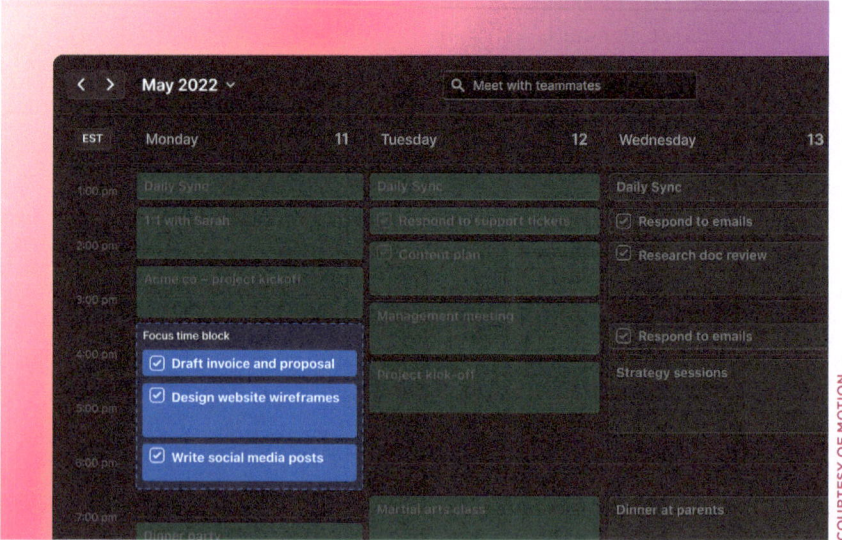

FIGURE 3.3
Automation tools like Motion can ease the friction of manual administrative work related to time management.

Tracking

Tracking tools help designers focus on a given task. They also help designers track the time they spend on project work. They can be useful for understanding activity trends, enabling teams to better estimate project work.

Options include:
- Clockify
- RescueTime
- Timely (Figure 3.4)
- Toggl Track

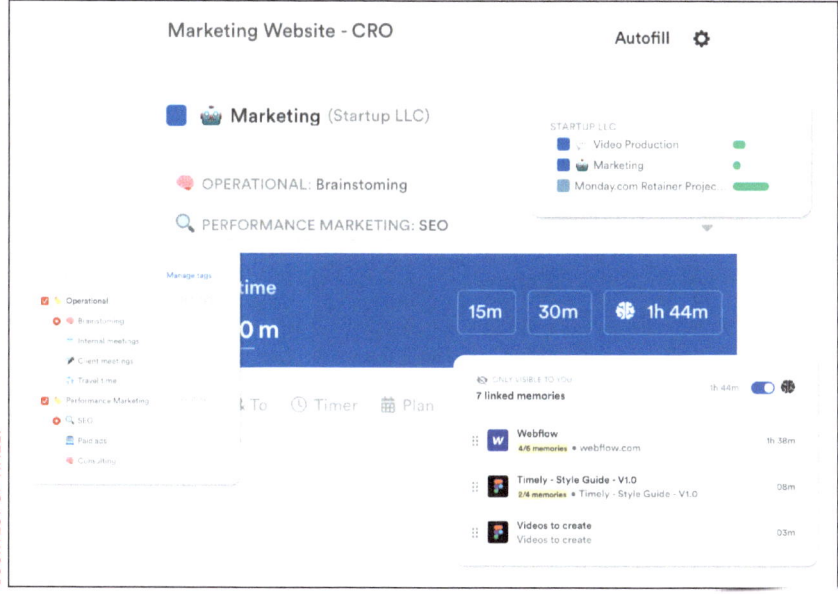

FIGURE 3.4
Tracking tools like Timely help designers improve at estimating their work by allowing them to understand how long a task takes.

Stop Context Switching

Within each day, a designer has a limited number of hours to make an impact. Many designers spend the majority of their hours on the job in sporadic meetings with occasional 30-minute stretches of time for focus work. During those short stretches, they attempt to eke out high-quality designs. As you can see in Figure 3.5, this is not sustainable.

FIGURE 3.5
A schedule with lots of context switching leads to lots of exhaustion. Notice the excessive number of meetings in pink.

According to a study conducted by researchers at UC Irvine and Humboldt University, it can take an average of just over 23 minutes to refocus after experiencing an interruption, which is known as *context switching*. Continuous interruptions have a harmful effect by producing "a higher workload, more stress, higher frustration, more time pressure, and effort." They make work take longer, as workers must take extra time to return to the original task.

Context switching by way of sporadic meetings makes it harder to deliver good work. The constant ramp-up and ramp-down is immensely disruptive. While you can make up for the reorientation time by working more intensely, the outputs are generally lower in quality.

I've observed that designers need daily blocks of solid focus time containing at least two contiguous hours to create high-quality designs. Instead of having intermittent meetings, many high-productivity individuals chunk their conversations together into multiple-hour blocks with 5–10 minutes between each meeting, which leaves large blocks of unused time left over to focus on later. See Figure 3.6 for an example of a schedule that is optimized for focus time.

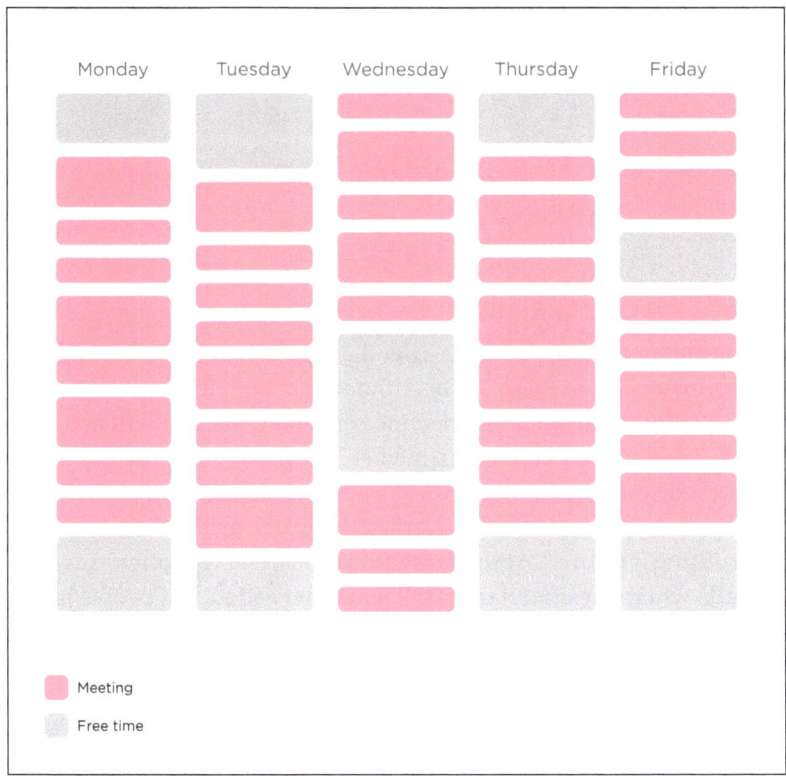

FIGURE 3.6
A schedule with 60+ minute blocks of meeting-free time gives designers the proper headspace to explore creative solutions.

This approach usually opens up several hours of heads-down focus time that can be placed either at the beginning or end of the day. During this time, designers can close their internal messaging applications and dive deep into the highest-priority work.

> **TIP** **HANDLE CONTEXT SWITCHING MOMENTS**
>
> If you can't avoid an upcoming context-switching block of 15–30 minutes, do not use that time for focused project work. You'll either barely make anything of value during that time or get into the design zone and miss your next meeting. Instead, use that time to do the following tasks:
>
> - Manage your personal task backlog.
> - Review meeting notes.
> - Clean up your calendar.
> - Respond to messages.
> - Clear your email inbox.
> - Grab a snack.
> - Get some coffee or tea.
> - Take a short walk to clear your mind.
>
> Each of these will enable you to show up to your next meeting on time and energized.

Audit Each Meeting

Staff designers optimize their calendar for daily impact. Instead of accepting every meeting invite without question, they review meeting agendas in advance and decline conversations that will not add anything of value to them. Justine Lee, Principal Designer at Visa, actively declines meetings if there's no clear or specific reason for her to join. "Some places I've worked default to an 'invite everyone to everything' culture." At these organizations, she was often "invited to calls where my attendance isn't actually needed, but people just autopilot invite me." By being picky about the meetings she attends, she carves out more space for herself.

A great staff designer will also reschedule meetings that are placed at inconvenient times. People's energy shifts throughout the day. Designers who identify and accommodate their energy patterns can optimize their meeting times to ensure that they bring their best selves to conversations. Introverts who feel more alert in the morning may want to participate in meetings before noon and then have focus time in the afternoon. Extroverts who are foggy when they first log in might want to schedule all their focus time in the morning and then meet with people after lunch.

> **TIP SET TEAM AGREEMENTS**
>
> Staff designers don't work in a vacuum. Designers embedded within a team can speak with their cross-functional partners and agree on communal meeting hours that best support everyone. This will result in more ideal, predictable meeting rituals optimized for teamwide impact.

Color-Code Calendar Blocks

Treating one's calendar as a design project can deepen the understanding of energy levels at a glance. Lara Hogan, former VP of Engineering at Kickstarter and leadership coach, recommends color-coding calendar blocks to visualize trends. Any color can have any meaning, but my tried-and-true color system is the one shown in Table 3.1.

TABLE 3.1 CALENDAR BLOCK COLORS

Color	Meeting type
Blue	One-on-ones
Red	Update-style meetings such as all hands
Green	Ideation and decision-making meetings
Yellow	Nonwork time such as commuting, eating, or a personal appointment
Purple	Focus time

Emojis are visual indicators that can communicate lots of information in a small package. They add even more meaning within a color system. Table 3.2 shows the emoji coding I add to calendar invites to show myself and others how I plan to use my time.

TABLE 3.2 CALENDAR BLOCK EMOJIS

Emoji	Meaning
⛔	Do not disturb—personal appointment or generally discouraged from booking
🎨	Focus time—booking discouraged if coupled with "do not disturb" sign
🍽️	Eating
🚆	Commuting time
☕	Coffee chat—social time

This trick illustrates trends within a week or month. It can, for example, show which types of meetings are draining and which result in more energy. If a designer feels exhausted after five back-to-back blue meetings, they might disperse this type of meeting throughout the week. If they feel energized after purple focus-time blocks, they might shift focus time to the morning so they can have synchronous calls in the afternoon.

Capture Your Workload

Once designers have a better outlook of their time, they can start to better manage and predict their workload. Capturing every existing task and lightly planning projects can help designers excel at managing their work without getting caught up in work about work. Like most design problems, investing a little time in this process can pay dividends by helping teams foresee and avoid potential process issues before they arise.

WORK MANAGEMENT TOOLS

There are a huge number of work management tools out there. Many companies provide teams with one or a few tools to manage their work. Designers don't have to choose what their team uses, but they can choose where to track their personal tasks. Below are a few examples of tools that allow employees to manage different kinds of work.

Personal Tasks Only

These tools can be used in tandem with team project tools to capture miscellaneous tasks that don't fit into project work:

- Any.do (Figure 3.7)
- Google Tasks
- Microsoft To Do
- TickTick
- Todoist

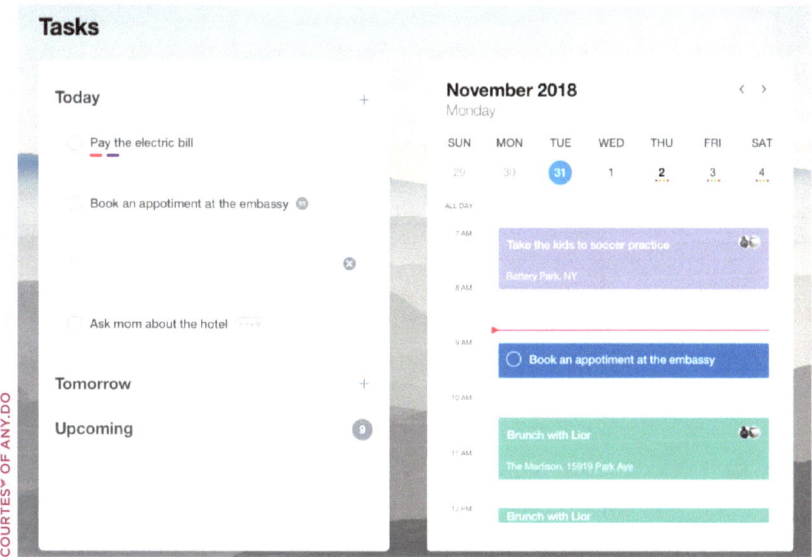

FIGURE 3.7
A personal task management tool like Any.do can help you keep track of the little tasks that live outside of projects.

continues

WORK MANAGEMENT TOOLS (continued)

Team Projects Only
These tools do not support personal work very well:
- Airtable
- Jira (Figure 3.8)
- Linear
- monday.com
- Smartsheet

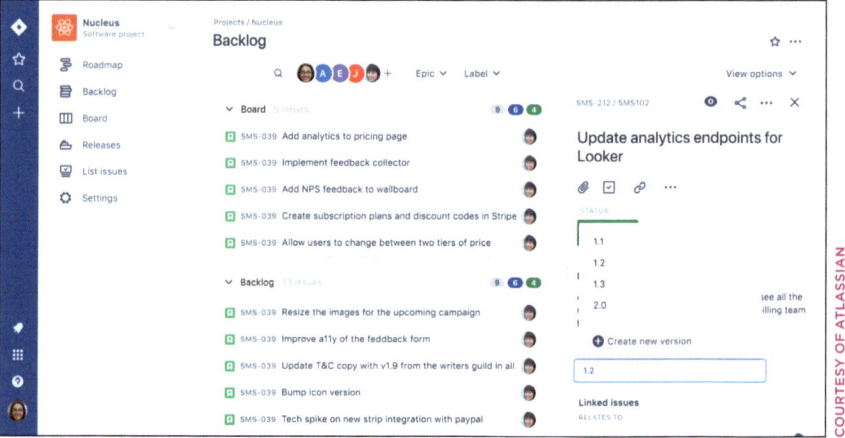

FIGURE 3.8
Tools like Jira help teams track and communicate about tasks related to their projects.

Personal Tasks and Team Projects

These tools can be used to capture both miscellaneous personal tasks and team project work:

- Asana
- Nifty
- Notion (Figure 3.9)
- Trello
- Wrike

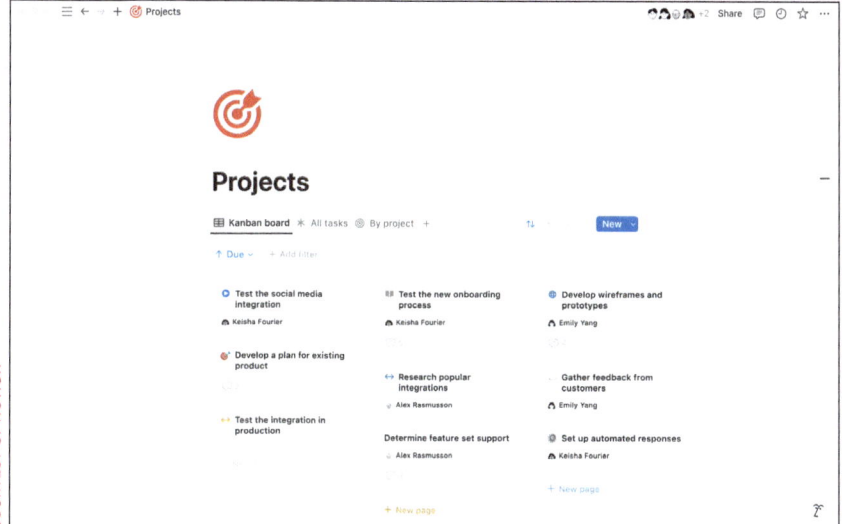

FIGURE 3.9
Notion allows users to create different spaces for personal and team tasks.

Build Your Backlog

Lots of designers have miscellaneous tasks that pile up through 1:1 meetings and team discussions, but they often don't document them anywhere. As you can see in Figure 3.10, all those tiny tasks add up. Therefore, they should be tracked along with project work.

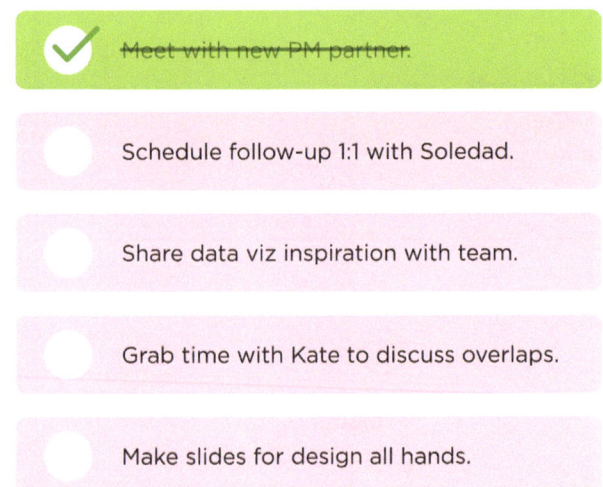

FIGURE 3.10
Miscellaneous tasks seem small in a vacuum, but they add up quickly.

Designers should have a backlog where they can quickly jot down the requests they have received from their manager, stakeholders, and collaborators in conversations. Examples of backlog tasks might include:

- Connect with an individual on another team to show them design ideas.
- Review a document that was shared with the team and provide feedback.
- Update slides based on feedback from leadership.
- Submit performance management assessment for a teammate.

Even if the tasks aren't documented and accounted for, others will expect them to get done somehow. Some of these tasks can be critical for building trust with senior leaders, which is key to increasing influence. It's important to make sure that these tasks don't fall through the cracks.

By listing out these tasks, designers can also visualize their workload more truthfully. They can name these tasks in conversations about capacity planning with the team, building transparency and empathy along the way. This ultimately results in better expectation setting across the team.

> **TIP** **TRACKING MISCELLANEOUS WORK**
>
> Designers often ask where to track miscellaneous work. This depends on the software your team uses to track their projects. Some tools make it easy to create private projects for miscellaneous work. Others are mainly built for collaborative projects.
>
> If your team uses a tool that isn't built to track personal work, you can absolutely use a different one. For example, my team used Jira when I worked at Etsy, but I tracked miscellaneous tasks in another tool called *Todoist* because that was easier for me. At Asana, we used Asana for individual and team work because it had support for both.

Plan Projects

Large team projects can overwhelm designers because they can be hard to plan. Sometimes, designers don't plan at all and instead jump into the work. This means they have no idea how long a project will take, and they figure it out as they go along. This puts designers at a disadvantage by preventing them from properly engaging with their team and setting clear expectations.

Project planning is an important strategic skill to have. Forecasting the steps required to complete a project can give designers the chance to negotiate with their team and senior leadership. They can agree on answers to questions such as:

- Should we run a design sprint?
- What kind of research does this project require?
- What devices and screen sizes should we design for?
- Which users and use cases are we designing for?

The answers to these questions can be documented in a short project brief like the one shown in Figure 3.11. Project briefs are usually owned by a project manager, but many teams do not have someone to fit into this role. As facilitators, designers can drive this conversation to get the answers they require to deliver their best work. By focusing the conversation on the questions that will impact design outcomes, designers can avoid overstepping boundaries and creating conflicts with product managers who often write product specifications or requirements documentation.

> ## Brief: New User Onboarding
>
> ### Objectives
>
> #### Our Hypothesis
> Because we know new users struggle to understand the core features, we believe creating a guided onboarding experience will result in increased user engagement and retention.
>
> #### Goals
> We will know this project is successful when:
>
> - 10% increase in user retention after the first week.
> - Positive feedback from user surveys on the onboarding process.
> - Reduction in support tickets related to basic feature understanding.
>
> #### Outcomes
> By the end of this project, we will deliver:
>
> - Proposed information architecture for the onboarding flow.

FIGURE 3.11
An example of a design project brief.

Designers who plan projects with their team make their work more visible, allowing teammates to better understand the design process. These designers also build a better picture of how long their projects will actually take. This awareness enables them to predict the steps necessary to create successful outcomes, and it sets them up to scale their work to other designers.

Set Healthy Boundaries

Designers and their teammates all have the same goal: to ship great work. A staff designer drives opinionated prioritization that contributes to a sustainable work culture. While a team can occasionally burn the midnight oil for an urgent launch, they can't continuously work in that fashion, or they'll burn out.

Sustainable work practices are upheld by sturdy boundaries. Designers can assert their boundaries by estimating upcoming project work, triaging tasks before accepting them, and learning to say "no" more often. These three methods turn designers from subservient people pleasers into impact-driven leaders.

Estimate Capacity

A major way to set boundaries is by going through the process of estimating design work. When designers estimate their work, they become more strategic partners to their teammates. Openly communicating the time necessary to complete design work also clarifies the process and shows that it's not just about making things pretty.

While it's impossible to be 100% accurate, a small amount of estimation work can result in a lot of value for designers and their teammates. Similar to sprint planning, reviewing and breaking down the work necessary to complete a design project gives designers a roadmap. The conversation about this roadmap can buy them the time necessary to create high-quality designs by encouraging a discussion about which parts of the process require more or less investment.

Estimation conversations can happen through many lenses, including the following:

- **Risk:** How much impact would a failed solution have?
- **Complexity:** How much effort would a successful solution require?
- **Impact:** What effect the solution will have on the business and its customers?
- **Reversibility:** How easy would it be to undo the ideal solution?
- **Priority:** How important is this work to the business?

Teams can choose the ones that work best for their context and then break project work down into either phases or design process milestones depending on scale. Using the above lenses, teams can evaluate each phase or milestone to converge on an approximate project size and delivery date.

Estimates are almost always wrong, but they can help teams focus and have necessary hard conversations up-front. When combined with time management tools, project estimates can create breathing room for designers to explore and land on higher-quality outcomes. The estimation process can highlight areas where the team might need additional support, enabling them to proactively acquire more help that will guarantee the success of the project.

ESTIMATE HONESTLY

Never underestimate the time you need, even if you have to disappoint people. Also avoid overinflating estimates to buy yourself more time. Both of these acts deplete trust. Instead, ask the questions you need to be as truthful as possible and negotiate with your team to get the outcome you need.

A common framework some product teams use is called the *Double Diamond* (Figure 3.12). Popularized by the British Design Council in 2005, this framework goes in and out of favor with the design community but can be helpful when discussing the stages of an upcoming initiative. The first diamond focuses on identifying the right problem to solve while the second one centers around developing and delivering an excellent solution. Each diamond involves a divergent and convergent period, enabling teams to explore opportunities before aligning on next steps.

Designers can tie phases of the diamond to activities and artifacts they plan to deliver as part of the project, thereby giving teams more insight into the thought behind the pixels. The amount of time required at each phase depends on the risk and complexity of the problem at hand. It might take one to several weeks to discover a problem space and converge on the right areas around which to explore solutions. Then it might take the same or more time to develop and deliver a solution.

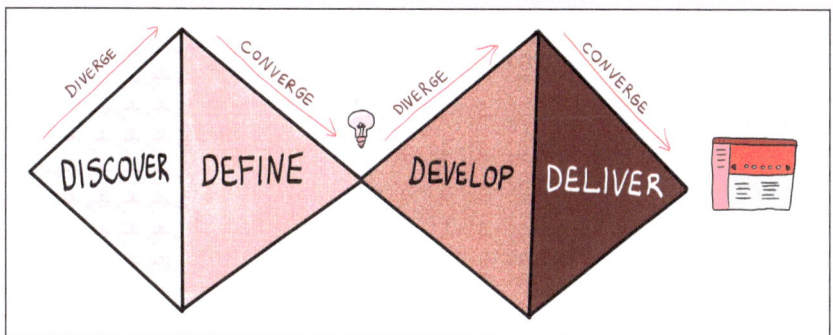

FIGURE 3.12
A framework like the Double Diamond can help you break down the work involved in developing an impactful solution.

Estimates must also account for the time it will take to get alignment from collaborators and stakeholders. Teams often forget to include design reviews and other buy-in approaches in their timelines, but each activity can add a week or more of preparation to the project plan. By proactively documenting the necessary advocacy work involved with getting a project shipped, you can add appropriate amounts of buffer time without raising eyebrows from your team and diluting trust.

Triage Tasks

Staff designers create a habit of assessing tasks before accepting them. Sunnie Sang suggested asking two clarifying questions first:

- Is it urgent, or can it be done later?
- Can anyone else do this work?

Designers can also ask for the priority of a task compared to their current project work. These questions help designers prioritize work with their teammates. Ultimately, triaging conversations can sometimes result in delegation or deprioritization of a request.

Triaging and organizing work based on impact is especially helpful for designers who receive lots of small requests. Plotting the requests on a matrix (such as the one in Figure 3.13) or in a spreadsheet can show teams which tasks should actually be prioritized. If a task is low-impact, teams can agree to either put it on a backlog or deprioritize it completely.

FIGURE 3.13
A prioritization matrix exercise can help designers work with teams to identify the work that matters most.

This process allows designers to build agreement to focus on more urgent work while also setting expectations about the kind of work they will take on in the future. At scale, designers and their teams can also use this process to prioritize projects on a quarterly basis. This results in better outcomes for designers, teams, and their customers.

CATT'S CORNER
Buy Yourself Time

When I began to take on larger scopes of work, it took me a while to figure out how to effectively plan and balance my workload. Even though I was tasked with strategic projects, my cross-functional team still expected me to support them with deliverables for smaller bug fixes and general experience improvements. They also expected me to find solutions to complex problems in the same amount of time as simpler ones.

At one point, my product manager and I both felt stretched thin, so we began to experiment with ways to organize our efforts so we could have more time for strategic thinking. One of my favorite lessons from those experiments is the importance of identifying which projects are less complex for design than engineering. If a team can prioritize a quick win that is easy to design but technically complex for engineers to build, that unlocks more time for the designer to deliver a great solution for a long-term, high-impact project (Figure 3.14). We began to stagger projects so that a long-term project followed a technically complex quick win.

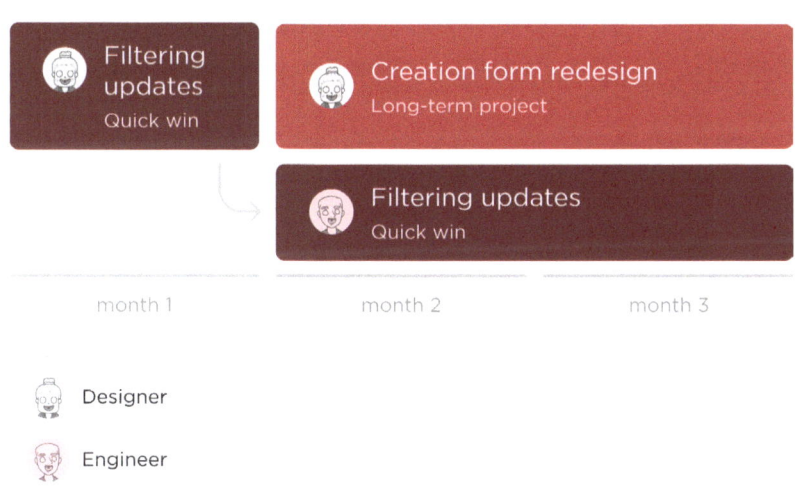

FIGURE 3.14
A solution that is simple to design but challenging to build can unlock more focus time for sticky design challenges.

When you are in a pinch and this technique doesn't give you enough time, there are still other ways to proceed before giving into pressure and sacrificing experience quality. I recommend conversing with the team about opportunities to pay down technical debt. Codebases always need maintenance.

If it's truly impossible to buy yourself extra time through other means, a short workshop can help identify opportunities to unblock engineers by separating the foundational efforts from visual design polish (Table 3.3). For example, perhaps the team can begin to build the skeleton of the experience based on the functionality listed in the product documentation and worry about the visual layer later.

TABLE 3.3 EXAMPLE OPPORTUNITIES TO UNBLOCK ENGINEERS

Skeleton Functionality	Visual Polish
Relationship between text string and user state	Treatment of text string
Button to submit filters	Button color and location
List of search results	Treatment of list of results
Elements included in a search result	Treatment of a search result
Transition between list and single page view of item	Treatment of item page

These are all ways in which you can maintain momentum while encouraging your team to respect your time. Never feel rushed by pressure from your engineering team. The more you negotiate and find solutions that work for all parties, the more your team will respect the design practice as equal to other functions. ∎

Say "No" More

Making people happy is a major reason many individuals become designers. However, saying "yes" to everything results in doing nothing well. Therefore, designers must learn to decline some requests so they can accept the ones that uniquely match their abilities and intended career direction.

Saying "no" to a request can be an important opportunity for the team to recenter what matters. Designers often get asked to do things that ultimately aren't that impactful and feel like they have to agree.

By triaging these requests, designers can figure out if the work is actually worth doing.

During my time at Etsy, I regularly overloaded myself with multiple low-priority projects for different working groups. Kristina Pyton, my design manager at the time, told me that I should fill up to 80% of my capacity with project work so I didn't burn out. I was instructed to spend the remaining time on skill-building, mentorship, and other forms of professional development. This would give me the space to stretch and fill 100% of my capacity when the business had sudden demands. I took her words to heart; since then, I've learned to better prioritize my work.

Declining work due to maxed-out bandwidth can also be a helpful signal. If you are unavailable to complete the work or need to delegate certain tasks to others, management must consider hiring additional support. I'll cover delegation more in Chapter 7, "Scale Your Impact."

"No" is always an answer, especially when the request will not move the team closer to their goals. Focus and teamwork is key to shipping great things. Staff designers view every request as an opportunity to build empathy and help their coworkers concentrate on what truly matters.

> **TIP** STANDING UP FOR YOURSELF
>
> Communication is a core competency as a staff designer. Many designers assume that protecting boundaries means communicating defensively; however, being too direct or harsh can shut down a discussion. An improvisational mindset can help designers assert their needs while also encouraging further conversation.
>
> Try operating from a place of openness and asking questions to understand the needs of others before turning them down. You may find other solutions to the request that still respect your boundaries. At the same time, communicating openly inspires kindness and flexibility in others.
>
> Examples of ways you can kindly communicate your boundaries include:
>
> - I need a bit of time to review my workload. Can I consider this and get back to you by end of day?
> - I'm balancing two projects right now. Would we be able to shift the deadlines for one or both of them so I can support you?

- How does your idea compare in priority to that of my current work?
- Love the enthusiasm! Thoughts on timing?
- I added this idea to our backlog, let's prioritize it together!
- That's not one of my current priorities; should it be?

All of these questions ultimately encourage the open lines of communication you need to be successful in a staff design role.

Debrief

Staff designers receive more requests than they have time to complete. Some or many of these requests will be complex and time-consuming. A good staff designer is aware of their capacity and sets appropriate expectations with collaborators, enforcing their boundaries whenever necessary.

Using the many time management work tools at their disposal, staff designers build awareness of their own availability. They determine how much work they can take on at a given time. With this information in hand, they work with collaborators to prioritize project tasks and create a sustainable workload.

Time and work management processes combine to build empathy for the effort that goes into creating great designs. It helps designers and teams deliver projects more predictably and with higher-quality outputs. Over time, these practices incrementally shift designers from being reactive to proactive, making them more strategic partners to their product management and engineering peers.

Activity

Calculate your average capacity per week and quarter.

Number of weeks in a quarter	Average hours of meetings per week *(Look at your calendar!)*	Average amount of focus hours per week *(40 minus meeting hours)*	Ratio of focus hours per week *(Focus time / 40)*	Amount of focus weeks in a quarter *(Weekly focus hours * 13)*
13				

What, if anything, surprised you about your capacity?

Plot 3–5 upcoming projects on a prioritization matrix and then reflect on the following questions.

What, if anything, surprised you about your prioritization?

What are 1-3 steps you can take to improve your situation?

Do you need to make some changes to your schedule so you can claw back some focus time? Alert your team that you'll be experimenting with some new time management techniques.

CHAPTER 4

Nurture Your Relationships

Build Your Network	78
Document Your Findings	86
Debrief	97
Activity	98

The staff designer role is highly collaborative, and relationships are key to success. Your network determines how easily you can make an impact through others. This means a well-connected staff designer has much more influence than an isolated one.

All relationships aren't created equal. When I was growing from senior to staff, I learned this the hard way. While I had wonderful relationships with my peers, I didn't have great connections with senior managers. This lack of connection led to missed opportunities. It also meant I was less visible than my peers, even though I contributed similar amounts of work to the company.

Building a network requires intentionality and focus. By proactively connecting with a variety of people from across the organization, designers can learn valuable insights, identify pervasive stories, and map out sentiment that might become relevant during project discussions. Using their artifact creation skills, designers can synthesize their findings into artifacts that they can personally reference later.

Build Your Network

Staff designers are expected to make big things happen without any direct authority. This is only made possible by building a solid network. Through strategic, regular points of connection with key individuals across the company, staff designers build a web of people with whom they can effect major change.

There are several categories of people a staff designer can connect with as they build their network (Figure 4.1).

- **Contributor:** Teammates and design peers who directly contribute to a project's outputs
- **Observer:** Minor stakeholders on neighboring teams who need to be aware of a project's progress
- **Approver:** Major stakeholders such as senior leaders who can either green-light or block a project launch
- **Supporter:** Cool folks who don't directly impact a project but are fun to speak with

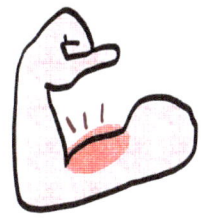   

CONTRIBUTORS   OBSERVERS   APPROVERS   SUPPORTERS

FIGURE 4.1
Staff designers interact with four categories of people in the workplace: contributors, observers, approvers, and supporters.

All these individuals are worth building relationships with through semi-regular conversations. Contributors can bolster ideas and expose cross-functional dependencies. Observers proactively flag concerns. Approvers and supporters can make connections between information in unique ways. These actions all assist a staff designer on their journey as an individual contributor who is also a senior leader.

Contributors

Teammates and peers who will be part of a staff designer's daily collaborative efforts are known as *contributors*. These individuals usually occupy roles such as a product manager, tech lead, engineer, UX researcher, or designer. Contributors directly impact the work a staff designer leads with their ideas and their individual outputs.

Cadence

Great relationships with contributors will make a project move more quickly. The more proactively a staff designer connects and shares work with these individuals, the better their own work will become. Lil Chen, Staff Product Designer at Discord, schedules one-to-ones with her closest teammates on a recurring basis and keeps a running notes document. She also sends her team leads direct messages so they "all feel safe speaking candidly to each other, especially because we're kind of running that team." These active efforts strengthen the relationships she has with her project contributors.

IN THE REAL WORLD

JESSICA HARLLEE

Jessica Harllee is a Staff Product Designer at Shopify. She and I met during our time at Etsy, where she worked for nearly six years. Between her time at Shopify and Etsy, she also worked as a Principal Product designer at Primary and a Staff Product Designer at Faire.

Throughout her 15-year career, Jessica has intentionally avoided becoming a manager. Back in 2019, she told me she wanted to keep doing hands-on design work because it best matched her interests. In May 2024, she began working at Shopify.

"Shopify is the biggest company that I've worked at. I've learned that because it's a big company, you have to quickly make connections. Figure out who to talk to early so that when you do need something or you have a question, it will be much faster to get answers. A big focus of mine was making connections, even if it was a 15-minute conversation just to introduce myself."

Jessica recommends bringing curiosity as you build relationships: "it's better to ask a lot of questions, and to maybe feel annoying"—especially if you're new to a company. The power of a staff designer role is often amplified by who they know. Jessica believes the goal of building a network is to "quickly figure out who to talk to and who will help you get something answered." The more people she meets, the better job she can do.

Examples of questions Jessica asks new collaborators include:

- What are some of the challenges that the team is currently facing?
- What do they think about the work coming up in the next year?
- How do they feel about certain observations of common behaviors or sentiments at the company?
- What are the qualities of people who do well at this company?
- What do you like and dislike about the way that we work today, and what are ways you hope that it can shift?

Jessica recommends aligning with your manager and design leadership on the areas that you and your team might want to push on in the future. The more you know their intentions, the more you can influence them and make an impact on your area of ownership. "It's not just one person at the top saying something. The team is being shaped from multiple angles. You're getting signals from both VPs and other ICs."

NOTETAKING TOOLS

As you meet with different people, make sure to take notes so you can synthesize what you learn later. There are lots of tools for taking notes.

Options for notetaking include:

- Pen and paper (Figure 4.2)
- Collaborative word processing tools, such as Google Docs and Notion
- AI notetaking tools, such as OtterPilot and Zoom AI Companion

Each option has its own benefits! Pen and paper is best for recalling information; word processing tools let you collaboratively prepare an agenda in advance; and AI tools let you focus on the discussion at hand. Choose the one that works best for your shared context so you can have the most organic conversation possible.

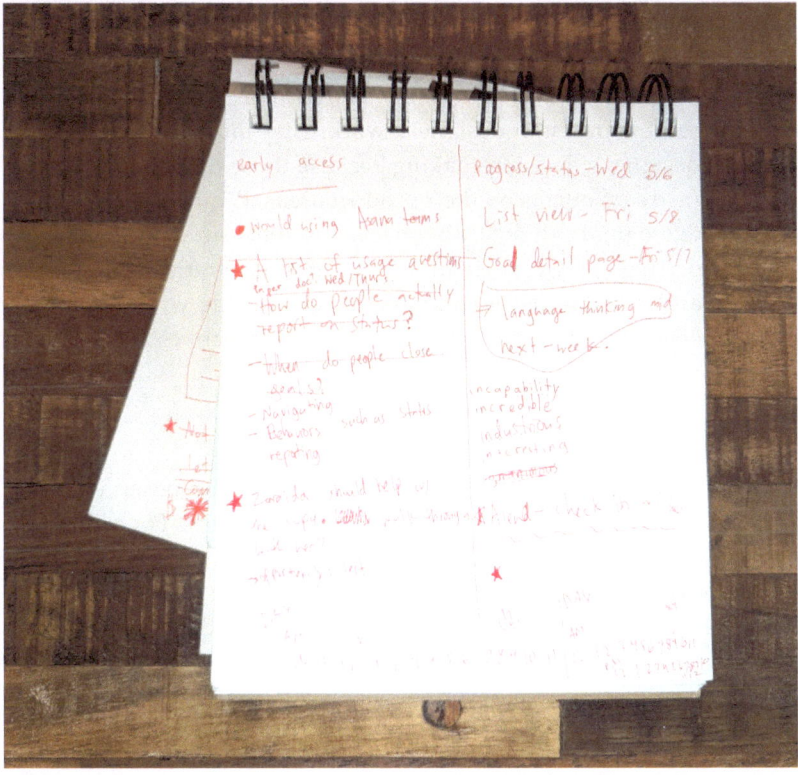

FIGURE 4.2
Pen and paper notes from one of my many notebooks.

Regularity promotes camaraderie, which ultimately breaks down silos and ensures that projects run smoothly. Staff designers should connect asynchronously with contributors daily through their company's messaging platform and their team's project management tool. Synchronously, key contributors should be consulted on a weekly or biweekly basis while lower-level contributors can be consulted as little as once per month.

Goals

Contributors likely have similar goals to that of a staff designer. However, they also probably have personal growth goals that can be bolstered by a collaborative relationship. Perhaps they are seeking a promotion or a raise and could use an advocate. Learning what matters to these individuals can result in both a deeper connection and fuel for the team's motivation.

Observers

Observers are members of teams who own product areas that intersect with a staff designer's work. They must be informed of the designer's decision-making because those decisions will have downstream effects on their product areas. While observers don't make the final call regarding decisions, they can still block launches by alerting approvers about the lack of consideration for their product areas. A staff designer shifts the relationship with observers from adversary to advocate by inviting their perspectives and including their needs at an early stage.

Cadence

A variety of ad hoc and scheduled meetings with observers help staff designers connect the dots on a regular basis. For example, a designer might facilitate an ideation workshop at the beginning of a project and invite several members of a partner team to participate. Then they might share weekly updates within the company's communication tool so observers can provide feedback. Members of their teams might meet in group settings, such as design critique, and otherwise connect on a monthly cadence.

WORKING AGREEMENTS

Contributors are your direct partners, and coordinated relationships with them can speed up project work. A working agreement with contributors will strengthen your mutual relationship. The simplest method to facilitate a relationship is to sit down with a teammate and list out your preferences for collaboration. Or you might organize a more elaborate workshop, complete with a workspace where everyone can visualize their working styles. Regardless, a set of working agreements (Figure 4.3) enables you and your contributors to agree on how you'll work together. You can document these preferences for yourself in what is referred to as a *working manual* (Figure 4.4). Example subjects to cover might include meeting hours, feedback communication preferences, and how you like to work with people in roles such as product management. Sharing these kinds of preferences up-front can help you avoid the friction that often comes with joining a new team.

How We'll Work Together

- Design owns visual and interaction design decisions.
- PM owns business decisions.
- Eng owns code and architectural decisions.
- Cancel any meetings without an agenda.
- Share updates at the end of every day.
- Raise red flags ASAP.
- Don't reopen closed doors unless it's blocking another decision.

FIGURE 4.3
Working agreements help product teams collaborate more effectively by giving them the opportunity to share expectations up-front.

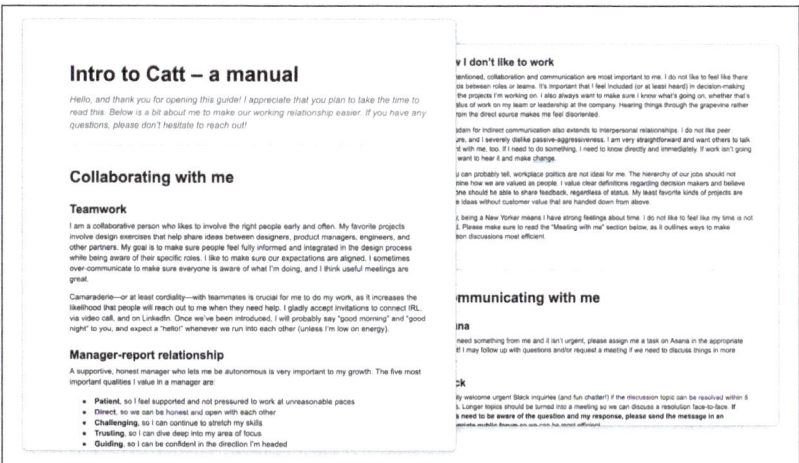

FIGURE 4.4
Working manuals help workers share their preferences with a wide array of potential collaborators.

Goals

People on partner teams likely have unique goals and intentions. It's important for staff designers to identify the overlaps and conflicts between team outcomes. This will allow teams to proactively communicate potential challenges to project approvers and resolve potential issues.

> **TIP NAVIGATING BLOCKERS**
>
> Even if you communicate regularly with observers, there still might be times where disagreements arise. If an observer raises a flag that might block your project, don't fret or become defensive! Instead, listen openly to their feedback and look for the areas of overlap between your goals. By concentrating on your shared ideal outcomes, you can find a solution that supports both your needs and theirs.

Approvers

Individuals who can make or break a team's project with a single "yes" or "no" are known as *approvers*. These folks are usually senior leaders, and they directly determine whether or not a project can proceed as planned. Due to their enormous power, approvers' opinions must be carefully considered in tandem with the needs of a product's customers.

Cadence

Approvers are usually busy, but their words can have a major effect on the outcome of a team's work. Therefore, staff designers must connect with project approvers early to provide them with up-front opportunities to provide direction. This can be done by scheduling a project kickoff or design workshop.

Throughout a project, approvers must also receive semi-regular updates. These updates will give them the opportunity to course-correct the team and give the green light when necessary. Staff designers usually meet with approvers on at least a biweekly basis to present the latest progress, which helps to get regular affirmative feedback and spot issues before they become toxic.

Goals

Members of senior management usually have goals that are connected to business outcomes. Staff designers can connect their efforts to approvers' priorities by learning what matters to each individual through stakeholder interviews. This approach will ensure that the team makes a larger impact.

Learning what makes your approvers tick will also give the team the language necessary to successfully advocate for projects they care about. This is because translating a project proposal through the lens of company goals is more convincing to a senior than an outcome that only benefits an individual team. Much more information about the value of this communication work lives in Chapter 6, "Build Influence Without Authority," which is all about building influence.

> **TIP HANDLING HIPPOS**
>
> Approvers, like all people, come with varying levels of ego. While many approvers are open-minded and collaborative, some are more authoritarian. Handle these particular individuals with care.
>
> If an approver tries to micromanage or block your team's work because they feel a personal need to control the outputs, create space for them to air their feedback. Look for opportunities to steer the conversation toward customer insights and shared ideal outcomes whenever possible. But also, be willing to accept that sometimes the HiPPOs (highest paid person's opinion) will win.
>
> In situations where HiPPOs tip the scale, make sure to document the outcome. Documenting these kinds of decisions can help the team keep track of all the ways a project transforms. This documentation can also serve as receipts if things go haywire and a high-ego approver tries to place the blame on your team.

Supporters

Some individuals have a great aura but don't make a direct impact on project work. These are *supporters*—they build a designer's confidence, share the same values, and might be great advocates one day in the future. Building a roster of these individuals can result in surprising impact because they enable staff designers to gather snippets of information from across the organization.

Cadence

Since supporters are the least impactful category of the bunch, a staff designer acts accordingly by meeting with these folks on a less regular basis. Quarterly meetings will suffice for most individuals; a select few might get the benefit of a shorter cadence, such as monthly. These chats are usually coffee-style conversations with no planned agenda. The expected outcome of discussions with supporters is usually to learn, listen, and share whatever is on your collective minds.

Goals

Supporters will have varying goals that may or not be relevant to a staff designer's work. As a designer builds relationships with their supporters, beneficial overlaps and collaboration opportunities may arise. The more conversations that happen, the more that opportunistic connections will be made.

> **TIP** **AVOID OVERCOMMITTING**
>
> Strong relationships with supporters can be beneficial in lots of ways. One of the best outcomes is that you might uncover new opportunities for your teams to collaborate. Don't bite off more than you can chew. Before you agree to any new work, connect with your team and prioritize potential opportunities by comparing them to your prior commitments. Also, alert your manager to confirm that no one else is already tackling the issue. If the opportunity isn't a priority right now, that's okay!

Document Your Findings

Designers often take the raw notes from their research and synthesize takeaways. Why not do the same thing with conversations at work? There are many ways to glean insights from the various discussions with coworkers. As shown in Figure 4.5, four important insights that can be gained from synthesizing discussions are:

- **Dynamics:** The relationships of power that exist between teams and individuals
- **Opportunities:** Potential new initiatives that might result in rewards
- **Sentiment:** How people feel about a topic
- **Symptoms:** Signals and stories that are repeated by multiple individuals

Each of these insights will help a staff designer learn to better navigate their organization.

FIGURE 4.5
Conversations with coworkers can help you uncover power dynamics, opportunities, sentiment, and symptoms.

Dynamics

Dynamics are the relationships of power that exist between individuals and teams. Power dynamics are important because they might be the difference between a project failing or succeeding. In a company, an individual can have high, medium, or low power. Ideally, a staff designer is supported by high-power individuals as often as possible. However, many leaders have moderate or low power.

High Power

Individuals who often seem to get their way have a lot of power and influence at the organization. High-power individuals are usually senior leaders, and they are likely responsible for approving project launches. This power was likely built up over a number of years, and this person is either trusted or feared by many. Sometimes, high-power individuals are new hires in major roles, such as VP, who inherit a heavy amount of influence. These individuals are crucial to the success of a staff designer.

People with lots of power must have their opinions considered often. If they are ignored, they will have a major negative impact on the outcomes of a team's work. Whenever possible, staff designers must connect with high-power individuals to learn their sentiment about various topics and build symbiotic relationships. This will ensure that those with lots of influence become potential advocates.

> **TIP** **BALANCE YOUR CONNECTIONS**
>
> If you mostly connect with high-power individuals, you might be missing crucial information from those who make the work happen. Be sure to connect with people who are closer to the ground on a regular basis. They'll help you build awareness and spot new opportunities that senior managers might be missing.
>
> Conversely, when none of your best connections have high levels of power, you and your team are more vulnerable to the whims of senior management. If you tend to connect most often with people who don't have much influence, consider branching out. Build more relationships with folks who have moderate and high levels of power so you have all your bases covered.

Moderate Power

An inconsistent track record of influence likely means that someone has moderate power at a company. These folks likely have strong opinions but must be careful when they express them. They may occasionally succeed, but they also get overruled by those with more influence on a regular basis—they are at the middle of the pack and must act accordingly.

People with moderate levels of power are often lower-level managers or team leads. They frequently make mid-level decisions and must be informed about the status of project work, but tend to get blocked by individuals with more influence or authority at the company. It's important to identify these folks because they can be connected to those with more power.

While people with moderate power can't stop a project in its tracks, they can make an impact on the right initiatives. A staff designer builds connections with subject matter experts who have moderate levels of power. These folks will have access to spaces the designer doesn't, and they will support initiatives that intersect with their own goals. Because of their connections to those with more power, they may be able to support a team in convincing leadership in times of disagreement.

Low Power

People whose opinions and decisions have little to no effect on the organization have a low level of power. These individuals will not be able to remove major obstacles or approve work. Instead, they add value and insight to a designer's work.

Most of the time, people with low levels of power are contributors, supporters, and observers. They may be individual contributors or newer managers with little influence. Since these folks are closer to the work, they may be able to share relevant information about challenges and opportunities affecting their areas of the organization. This can help you operate more effectively.

> **POWER MAPPING**
>
> There are different ways you can document the levels of power that individuals have at a company. Iyobosa Irabor, founder and CEO of the Insightfully coaching practice, recommends that designers visualize the relationships of power between individuals to figure out how to position themselves in the workplace. "Map key players' goals, priorities, and struggles against organizational priorities. Observe persuasion patterns—who uses which influence channels, when, and how effectively?"
>
> By applying the same pattern recognition skills from hands-on design work to interpersonal relationships, Iyobosa believes designers can "understand exactly how successful initiatives gain traction, with whom to align, and the most effective approaches for advancing your own theories of success." Illustrating your ecosystem creates the same internal clarity your team feels when you create a user flow diagram for them. Two examples of valuable maps are the power ranking and stakeholder matrix.
>
> A power ranking (Figure 4.6) can help you understand who to connect with on a regular basis so your work can go as smoothly as possible. Start by listing out all the individuals you know and then rank them according to the amount of sway they have at the company. This may illustrate some gaps in your influence—perhaps you frequently meet with low-power folks and only connect with high-power individuals on a quarterly basis. Ideally, you should have a healthy mix of high-, moderate-, and low-powered folks on your regular schedule.
>
> *continues*

POWER MAPPING (continued)

Name	Org	Position	Power	Priorities
Alice	Design	VP of Design	High	• Consistency • Design quality
Prashant	Design	Director of Product Design	High	• Business strategy • Design quality • Product development process
Charlie	Design	Design Manager	Moderate	• Consistency • Professional development
Jiawen	Design	Senior Product Designer	Low	• Design quality • Design velocity • Product development process
Eve	Design	Product Designer	Low	• Design quality • Design velocity
Frank	Product	VP of Product	High	• Business strategy
Grace	Product	Group Product Manager	Moderate	• Business strategy • Product development process
Ivy	Product	Product Manager	Moderate	• Business strategy • Design velocity
Vivian	Engineering	VP of Engineering	High	• Business strategy • Code quality • Product development process
Katie	Engineering	Engineering Manager	Moderate	• Code quality • Design velocity • Product development process
Gordon	Engineering	Tech Lead	Moderate	• Code quality • Design velocity

FIGURE 4.6

A power ranking can help you identify the people who are most influential and reflect on what they care most about.

A stakeholder matrix (Figure 4.7), created by Aubrey L. Mendelow in 1991, can help you map and connect stakeholders' power to their level of interest in your project. Intersecting these two categories of information enables you to identify the best way to expend energy on efforts to get buy-in and support from approvers and other individuals you work with. For example, you'll want to work closely with high-power approvers who are very interested in your work but put in the bare minimum for low-power observers who couldn't care less.

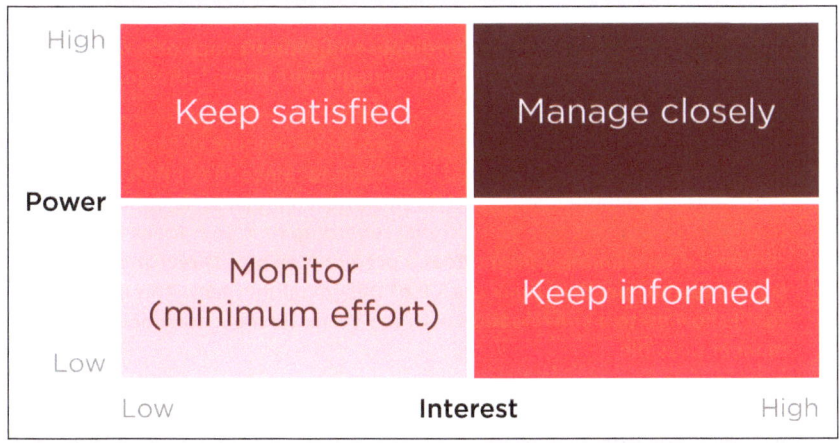

FIGURE 4.7

A stakeholder matrix can help you determine the people who should be given the minimum or maximum effort.

CATT'S CORNER
Contextual Communication

Networking takes energy, and we all have a limited amount to expend on socialization. I use my energy more effectively by shifting the way I speak based on the level of power held by the individual I am speaking with. Considering the potential impact of these conversations on my goals and the goals of those around me helps me adapt my discussion topics appropriately.

Individuals with less power won't be able to directly affect change, but they may become allies who can share relevant insights from their corner of the organization. When speaking with low-power individuals, I usually concentrate on information-sharing in both directions. By sharing observations with each other, we can coordinate and support each other as we both build influence. I usually speak quite casually with these individuals, as they are unlikely to be approvers.

When I speak with people who have a moderate amount of power, I shift between information-sharing and idea-sharing. Since they often have a direct line of communication with high-powered individuals, folks with moderate power can give you a signal regarding how your ideas will be received. For example, I might float a project idea by a Director of Product Design to see how they think the VP of Design might react. This ensures that I show up to a conversation with the VP in the most convincing manner possible.

High-powered individuals are well-connected and influential. These folks are busy, so I try to only bring the most relevant discussion topics their way. It's critical that these folks have a positive opinion of your work. You'll learn more about managing up and executive presence in Chapter 8, "Show Your Value." ■

Opportunities

Potential initiatives you and your collaborators might decide to take on are opportunities. Sometimes, opportunities arise from symptoms. Other times, individuals will mention opportunities during conversations. The potential to do meaningful work increases with the number and scale of opportunities that a team uncovers.

While there are endless opportunities to improve and make an impact, there is limited time. A staff designer takes note of opportunities and prioritizes them to decide which ones deserve their precious time. Some opportunities might be best handled by someone else; in that case, the designer can pass them along to the right parties as suggestions.

> **TIP** GIVE CREDIT WHERE IT'S DUE
>
> Make sure to credit the people who share opportunities with you. While ideas are cheap, attribution can have major effects. Crediting the original person who shared the idea with you helps build camaraderie and trust amongst coworkers. And if the outcome has a positive effect on the company, it could help the credited individual get a promotion.

Sentiment

Humans are full of feelings. And whether or not people want to admit it, work is a highly emotional experience. The way an individual or group feels about a particular topic is known as their *sentiment*.

Sentiment is important because it can directly determine how easily the work will happen. It can affect the speed at which work is delivered; an exciting project will move faster than a mundane one. And in the worst-case scenario, it might even determine if a team will be able to roll out a project.

Designers can learn the sentiment of collaborators, approvers, and observers by bringing up specific topics during conversations. People will usually be up-front about their feelings, especially if they are excited about a particular project. Some individuals might be cagier about their opinions, but they might reveal frustrations in private discussions. This is why it's important to connect with folks in both group and 1:1 settings.

The three types of sentiment are positive, neutral, and negative. Staff designers recognize the value of sentiment and adjust to accommodate it. Whenever possible, they look for projects that people appreciate and work to either neutralize negative opinions or avoid projects with negative sentiment.

Positive

When a person feels good about a topic or initiative, they have a *positive sentiment*. It's clear that someone feels positive about a topic if they respond affirmatively or even with excitement. Another way to detect positive sentiment is if an individual proactively brings up a subject and makes passionate statements in favor of it.

Whenever possible, look for topics that people feel positive about. When people respond well to a subject, they will likely help build influence in favor of it. These people are natural advocates for the work they care about. If a senior leader is passionate about designing quality experiences and a project might help the team improve, for example, tying the project to that initiative can result in a larger impact.

> **TIP STAY ON TOPIC**
>
> Positive sentiment can be a great resource for the creation of mutually beneficial outcomes. It might be tempting to try and connect an unrelated idea to a subject that someone is passionate about. However, this can backfire and result in decreased trust.
>
> If you try connecting a project to an idea just because someone feels positive about it, that can come off as ungenuine. Avoid attempting to shoehorn an unrelated project into a topic that has positive sentiment. Instead, look for authentic chances to link your work to topics that matter to people. These might sometimes feel rare, but the genuine nature of these connections is what makes them special!

Neutral

When someone isn't expressing excitement about a topic, but also isn't downplaying it, they have a neutral sentiment. A *neutral sentiment* means someone doesn't have intense feelings about the subject. Employees have neutral feelings about most work, so it won't be easy or particularly challenging to convince a teammate to do something they feel average about.

If someone feels neutral about a subject, that's okay. There are generally two ways to proceed. You can try to nudge them toward positive feelings by sharing relevant information that illustrates the value of your initiative. Or they might agree to follow the group if enough powerful people are on board. Peer pressure is a useful tool!

Negative

If someone expresses frustration or actively downplays the value of a specific topic, they have a negative sentiment. *Negative sentiments* imply a person has a misalignment of goals or wishes that the issue would no longer exist. These are subjects to either combat or avoid associating work with.

SENTIMENT MAPPING

A great way to assess sentiment regarding a topic is to map out the org chart and then illustrate what you know about the sentiment of your coworkers (Figure 4.8). This will show your supporters, potential blockers, and information gaps. All positives or neutrals would be an ideal outcome, whereas gaps or negatives might denote more concern.

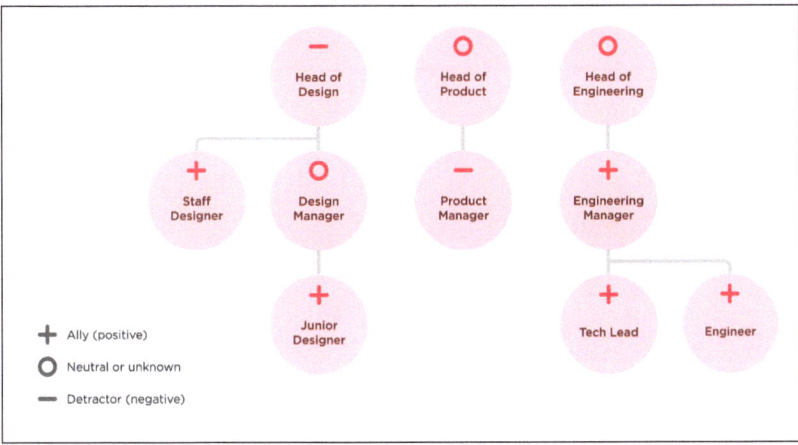

FIGURE 4.8

A sentiment map can help you understand why a project isn't getting the level of traction you hoped for.

For example, let's say that your manager is neutral about a project and the head of design feels negative about it. That might mean that your team could easily influence the manager. However, you will need to spend lots of time getting buy-in from the head of design as early in the process as possible.

Nurture Your Relationships   95

SYMPTOM AFFINITY MAPPING

Like user research, it's easy to spot symptoms when you make an affinity map. You can create an affinity map of symptoms (Figure 4.9) by taking all your meeting notes from a period of time—ideally a two-week period, a month at most—and looking for similarities. Highlight every story you heard during that period of time and compare them. Which ones were repeated most often? These are likely your biggest opportunity to make a major impact.

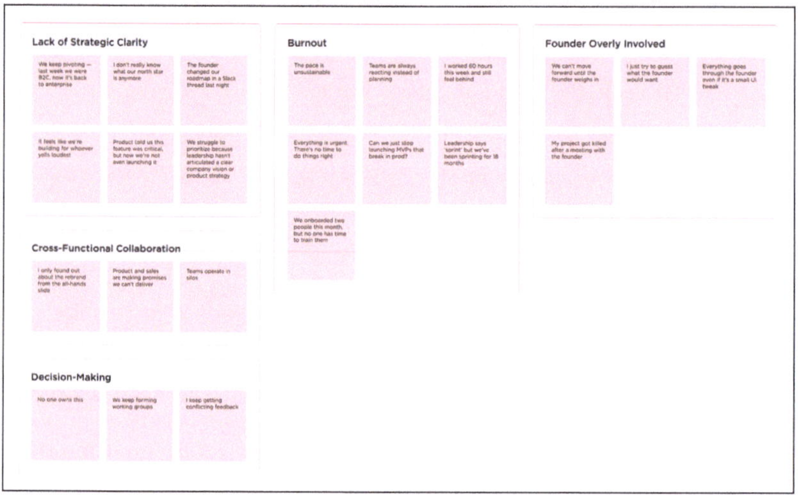

FIGURE 4.9
A symptom affinity map from conversations with employees at an example company.

Projects that elicit negative sentiment will face many challenges. The more senior the individual who is opposed to the work, the worse the experience will be. It's annoying if an unhappy engineer derails a project; it's tragic if a skeptical director can completely end it.

> **TIP RESPECT OTHERS' OPINIONS**
>
> If someone feels negatively about a subject, don't try to change their mind or sell them on your idea. After all, everyone has ideas! The goal is to create favorable outcomes for your customers and the business. Instead, listen to their feedback. They might have valuable perspectives that can make your work better.

> Edwin Morris, Staff Product Designer at Datadog, once shared a writeup describing an exciting new concept with staff engineers at the company. One engineer proceeded to comment with vehement disagreement because she had concerns related to rolling out the idea to users. Edwin reached out to meet with her live, and, according to him, "she had amazing feedback." Ultimately, Edwin pivoted the project, and, thanks to how well he handled her feedback, that engineer became one of the project's biggest champions.

Unanimous buy-in is the key to delivering a project on time. Staff designers ensure that their senior leaders all have positive or neutral sentiments regarding a project they plan to take on. If anyone expresses a negative viewpoint up-front, the issue can be proactively resolved. That's much better than someone outing themselves as a blocker midway, which would be a more challenging situation to recover from.

Symptoms

A story that has been repeated by multiple individuals at a company is a *symptom*. Symptoms are signs of larger problems at an organization. A symptom could be as small as a challenge that a specific team is facing or as large as silos that interfere with many teams.

Symptoms are especially important to identify when a staff designer is new to a role. By listing the stories they hear from different people, a new staff designer can locate the problems that are most pervasive. This can help them avoid falling into the same traps that their predecessors were hindered by.

Debrief

Relationships are key to becoming an influential staff designer. Cultivating a large and wide network of connections who directly and indirectly affect a team's projects can help designers mitigate potential concerns and identify future collaboration opportunities. As relationships deepen over time, insights related to sentiment, dynamics, symptoms, and opportunities will crystallize. These insights give designers the qualitative internal data necessary to prioritize their actions, ultimately helping them make the most impact.

Activity

Meet with key collaborators and stakeholders at your company and then document what you know about them.

Name Who are the key individuals in and around your work? List them out.	Role For each person, include their role and job title.	Impact How does this person affect your work?	Motivation What does this person care about most?
		Pick one: 💪 Contributor 👍 Approver ℹ️ Observer 💜 Supporter	
		Pick one: 💪 Contributor 👍 Approver ℹ️ Observer 💜 Supporter	
		Pick one: 💪 Contributor 👍 Approver ℹ️ Observer 💜 Supporter	
		Pick one: 💪 Contributor 👍 Approver ℹ️ Observer 💜 Supporter	
		Pick one: 💪 Contributor 👍 Approver ℹ️ Observer 💜 Supporter	

In the mind mapping tool of your choice, take your meeting notes and affinity map symptoms. Highlight any opportunities that came up in conversations; reframe any painful symptoms as opportunities, too. Discuss your findings with your team and manager.

CHAPTER 5

Drive Product Vision

Leaving a Seat at the Table	101
Take Back the Wheel	106
The Value of a Vision	111
Debrief	118
Activity	119

When a company hires or promotes someone into the staff designer role, they are placing a bet that the cost of the individual will be paid back in multiples. Staff designers deliver work at a pace and scale that most designers cannot, and they can generate a roadmap's worth of ideas in a small amount of time. Organizations with these roles are essentially investing in the future of the business.

Many of the staff designers I spoke to described vision work as a key expectation of their role. Theresa Slate, Staff Designer at Thrivent, spends "20 to 30% of my time" on envisioning the future. "I need visioning work in order to be successful. I can't deliver high-quality designs if I don't understand how our work ladders up to something bigger." Illustrations can be high-fidelity like the kind shown in Figure 5.1, or as low-fidelity as simple sketches. Visioning helps staff designers and their team align on how to allocate their time and energy. Thanks to her vision work, she and her team invest more time and effort into projects that are a strategic priority for the business.

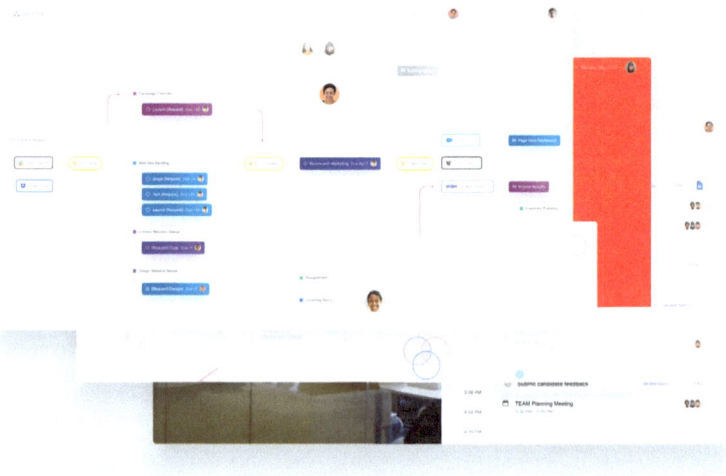

FIGURE 5.1
A screen from Asana's vision for the future of work, illustrated by their design team and shared publicly via YouTube in 2020.

The ability to define, prioritize, and deliver a large scope of future work is exactly what makes a staff designer so powerful. The shift from design execution to vision-setting is a learned skill. When senior designers get more involved in defining the roadmap for their product area, they can become strategic enough to make the leap to a staff design position. This moves the role of design from paintbrush to product partner.

Leaving a Seat at the Table

Over the past ten years, design leaders have written various pieces about lacking "a seat at the table." It's not a discussion about literal furniture; the "seat" refers to power and influence, and the "table" refers to the kinds of decision-making conversations that impact company strategy. This line of conversation is pervasive in our line of work.

The popularity of this phrase is so high because many companies employ leaders who do, in fact, exclude the design team from decision-making. Some companies employ executives who do not understand the value of design. This is especially common at large, older institutions with a long legacy. In these circumstances, designers actively try to contribute but are systematically shut out. It can be frustrating to navigate.

That said, many businesses believe in the power of design and view it as a superpower. Design is more powerful and necessary than ever for a company to succeed, especially as it scales. IDEO's Gemma Lord agrees—in March 2025, she reflected on this shift in a *Design Week* piece called "Designers Have a Seat at the Top Table—So What Now?"

> We are no longer just makers; we are facilitators of strategy. Systems thinking, problem-solving, and the ability to bring abstract ideas into tangible form is making us valuable beyond aesthetics. Designers are becoming integral to helping leaders and organizations make the right choices about their futures.

On a panel at ConveyUX 2025, Scott Lambridis, Nathan Shedroff, Debbie Levitt, and I reached the same conclusion. Design has actually become a strategic asset to businesses, and more leaders understand the value of critical thinking through the lens of design. In 2025, there is no shortage of opportunities for staff and principal design talent. But some designers and design leaders still make the mistake of backing away from contributing to business strategy because they think it's not their job.

Designers who decline to engage with business strategy are giving up their own agency. In these situations, designers are keeping themselves from being present at the table—or worse, walking away when the seat is offered. When designers learn the language of their cross-functional teammates and use design to support strategic thinking, everyone wins.

There are many moments during the design process where individuals give up the opportunity to proactively participate in the product development process. Sometimes folks intentionally avoid participation, and sometimes the opportunities are hard to identify. The four key themes of missed opportunities are:

- **Renounced power:** A designer declines to ask questions or otherwise participate in strategic conversations.
- **Lack of insight:** No qualitative research is used to scope or deliver a project.
- **Fidelity mismatch:** The fidelity of design assets does not reflect a designer's fidelity of thinking.
- **Perfectionism-driven silos:** A designer does not share work with collaborators until it is "ready."

All these actions subtract from the superpower of design work. They make designers look like paintbrushes rather than strategic partners. When designers focus less on delivering customer value and more on aesthetics, everyone loses.

Renounced Power

Designers can help teams proactively identify and resolve risks. This powerful skill should not go overlooked, nor should it be underutilized. When a designer does not give feedback or ask questions about a directive, they are renouncing their own power.

Giving up agency at the beginning of a project has harmful downstream effects. When a project is kicking off, a designer can choose to either ask curious questions or say little to nothing. Choosing to say nothing is an explicit action that subtracts power. It's like walking away from the table before the seat was offered.

Lack of Insight

Customer-aligned products are proven to have more successful outcomes in the market. Designers who do not employ qualitative user research techniques, such as concept and usability testing, rely on their guts to make design decisions. Without consulting their user base, they are setting themselves up for potential failure.

When a team is tasked with solving a complex problem, confidence in a solution can only come from a consistent feedback loop with customers. A lack of insights can be the difference between success and failure. Moving forward without the meaning-making that comes from gathering qualitative data can result in wasted energy, failed product launches, and a decreased trust in design as a strategic partner.

Fidelity Mismatch

The visual layer of a design can communicate its progress toward launch. Thanks to the many design tools on the market, designers can now quickly deliver designs that look very polished, even at the idea stage. Design systems make the exercise of delivering polished concepts more appealing, as ready-made high-fidelity components can be used for ideation.

While high-fidelity concepting feels like the team is moving faster, it can actually be harmful in the long run, depending on the culture of your company. People think the fidelity of an illustration expresses the fidelity of the thinking behind it. This can move the topic of discussion from the validity of the idea to the quality of the visual treatment. When an idea still needs more iteration, but the design looks complete, collaborators may interpret it as such. This subtracts the time necessary to find an ideal solution and conflates strategy with visual and interaction design.

> **TIP DESIGN CULTURE AFFECTS PERCEPTION**
>
> Fidelity mismatch is a serious challenge that some companies face; however, the level of this issue's pervasiveness may depend on the culture of your company. Leaders from companies like Anthropic, Netflix, and Notion told me that their organizations have a culture in which people effectively discuss ideas, even when looking at high-fidelity prototypes.

Randy Hunt, Head of Design at Notion, noted that having a founder who was a designer has resulted in a respect for the value of design as a differentiator. The company values momentum, and designers are encouraged to learn as quickly as possible. "We learn a lot by putting things we make in front of people. This doesn't mean we move fast and break things—in some cases, we get something running in a local environment, sit down with an important customer, and show it to them." This scrappy nature encourages a higher level of polish at the ideation stage.

The unique needs of your leadership team may also impact the way you treat design concepts. If your senior leaders struggle to grasp concepts without seeing them in a highly polished context, you will need to adapt appropriately. However, that doesn't mean you need to use the same assets when discussing concepts with your team. Conserve your creative energy whenever possible!

Perfectionism-Driven Silos

Perfectionism is rampant in the design field. Designers are often taught to continue tweaking their work ad infinitum until every part of the work is "ready" and every detail has been thoroughly considered. Individuals are discouraged from presenting anything that has not been thoroughly iterated upon, for fear of showing imperfections. This culture is reinforced by many senior design leaders, who pressure their reports to deliver the most beautiful and usable solution out there. Some say the goal is to push the industry itself forward.

Design is a unique skill set, and it can often feel isolating or frustrating to practice. For many individuals, owning the visual layer and communicating with partners only on an as-needed basis helps maintain a sense of control. Individualized control, which stems from our ego, ultimately hinders the user experience of a product.

Cultural pressure and the need for control often leads designers to constantly iterate on low-value work without communicating progress to their collaborators. Hiding work removes the opportunity for the team to give early directional feedback. People in other functions are responsible for the final definition of project scopes, so iterating without them will result in scope cuts, wasted energy, and diminished trust.

IN THE REAL WORLD

SARRAH FIGUEROA

Sarrah Figueroa is a Staff Product Designer at a major content publishing platform. We met while she was working in a similar role at an analytics software company called Heap. I immediately knew Sarrah was an incredible strategic thinker. She has a background in workshop facilitation and information architecture, which makes her a mighty partner to product and engineering.

Through curiosity, Sarrah identifies opportunities for closer collaboration with her cross-functional peers. "I usually come into a job, and they put me on a scrum team. But I'll ask questions to help them figure out what they are trying to do. And then they realize they actually need to figure out the strategy, so they ask me to help them work on it." She doesn't wait—she shows value and then the team begins to consult her more often.

Sarrah is all too familiar with the conversation about the seat at the table. "There are some companies that are more closed doors, and some companies that are more open." If she works with someone who won't invite her to conversations, she will proactively reach out and consult the individual to learn more. "History can tell you a lot. Maybe a more junior designer was brought into that room and didn't contribute or participate."

She recommends setting clear expectations with peers so they understand the value a designer can bring to the conversation and then following through with tangible impact. This is especially critical in situations where collaborators have had prior negative experiences. "What you have to make clear is that it's not a fear of missing out. There's a value-add. And then when you do show up in the room, you have to make sure that you are adding that value."

Take Back the Wheel

Awareness is the first step to reclaiming agency as a designer. The second step is taking action. Every interaction with a collaborator is an opportunity to move toward more active participation in the product development process. Four key ways to shift from reactive to proactive are:

- **Bring curiosity.** Step up and ask questions rather than accepting assignments.
- **Increase confidence.** Support strategic decision-making with customer insights.
- **Progressively upscale.** Ensure the fidelity of your outputs reflect the current state of your thinking.
- **Design in public.** Beat perfectionism by sharing works in progress.

These actions solidify designers as strategic partners instead of paintbrushes.

Power: Bring Curiosity

After receiving information about a new project, a designer might feel the temptation to get started without asking questions. It can be rewarding to move quickly when you're given a new project. But this reinforces the idea that you are amenable to taking orders without question.

A staff designer is a critical thinker who optimizes their effort for the most impact. Clarifying the intent behind a scope of work can help the team align on expected outcomes and have more confidence in the overall project directive. More importantly, a few simple questions might uncover risky information gaps before they become major blockers.

Examples of questions a staff designer might ask at the beginning of a project include:

- Who fits into our target audience?
- What are open questions that research can help us answer?
- Where else in the product might this experience provide value?
- Why are we planning to launch this experience on the intended date?
- Why do we believe this change must be made now?

As a designer, this kind of information ultimately affects the customer experience. If answers to any of these questions appear to be missing, the entire team has a right to know.

> **TIP REALLY BE CURIOUS (NOT CRITICAL)**
>
> When you decide to speak out, be aware of how you bring yourself to the conversation. Tone is really important. Some designers default to communicating in a hypercritical way. Meghan Logan, Staff Product Designer at Thrivent, believes this comes from a feeling of lacking ownership. "PMs are stereotypically seen as the owner of product impact, and engineers are seen as the owner of executing the experience. Designer contributions and impact get lost in the middle. That bubbles into resentment and comes out as defensiveness."
>
> If you act defensively, you will put other people on the defensive. The team won't be incentivized to include you if they feel like you won't be collaborative. Meghan recommends that designers learn to regulate their emotions and consider ways to approach encounters in a level-headed manner. "There's always a path forward, and you can't grow into a true leader who builds relationships and allies across an org if your ego is still present in those calls."
>
> The goal is to be supportive, not indignant. You do deserve to contribute strategically, and you do deserve accolades. But in order to get what you want, you'll need to remove negative emotions from your toolkit. Show that you genuinely want what's best for the team, and you'll be rewarded!

Insight: Increase Confidence

Once a project kicks off, a designer can choose whether or not to consult customers throughout the ideation and refinement phases of work. Seasoned designers have years of experience that do build a potent intuition. However, excluding real users from the design process can sometimes result in avoidable mistakes.

A staff designer will keep the customer at the forefront of the team's mind using artifacts like the one shown in Figure 5.2. They work to answer the biggest, most risky questions with the help of qualitative customer insights. They can help their product manager build a clear sense of the result that a particular effort might produce. Designers who are seen as partners in risk reduction get invited to decision-making conversations because they have a track record of effectively building confidence and launching successful solutions.

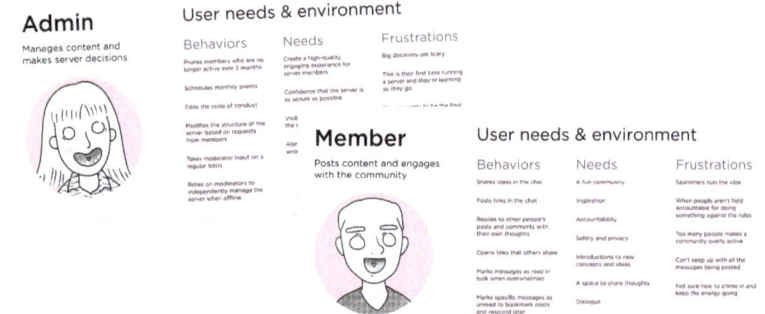

FIGURE 5.2
Artifacts such as personas can help ground your team in the types of customers they want to prioritize features for.

> **TIP QUICK AND DIRTY RESEARCH**
>
> Designers often receive pushback from cross-functional collaborators when they suggest research due to concerns about speed. However, a staff designer knows that good insights don't have to take forever. With the help of technology and some creativity, designers can help align the team on the way forward without spending weeks to prepare. Asynchronous user tests, card sorts, and surveys are three great methods to get customer feedback quickly.
>
> Research doesn't have to be large-scale or costly to be effective. In 2021, Raluca Budiu, Senior Director of Data Strategy at the Nielsen Norman Group, noted that research with five users can help identify major qualitative issues to resolve in an experience. That won't help your team build numerical confidence, but it can give you a list of problems to resolve. Check out *The User Experience Team of One*, 2nd edition, by Leah Buley and Joe Natoli for more examples of research techniques that help you do more with less.
>
> Unable to connect with actual users? You can lean on best practices, insights from research institutes, and guidance from subject matter experts. For example, growth designers might reference recommendations from "The Psychology of Design" by Growth.Design, a compendium of over 100 cognitive biases and principles that might impact a user experience.

Fidelity: Progressively Upscale

When a designer's work does not reflect fidelity of their thinking, it encourages the team to discuss visual design. Instead, they should be discussing the concepts behind the design. Staff designers recognize the power of visual fidelity and use the appropriate treatment to steer the conversation.

There are many ways to ensure that teammates recognize the fidelity of illustrated concepts, including:

- Hand-draw interface sketches (as shown in Figure 5.3).
- Create wireframes in a whiteboarding tool.
- Use a low-fidelity design system theme.
- Apply a color overlay to mock-ups.

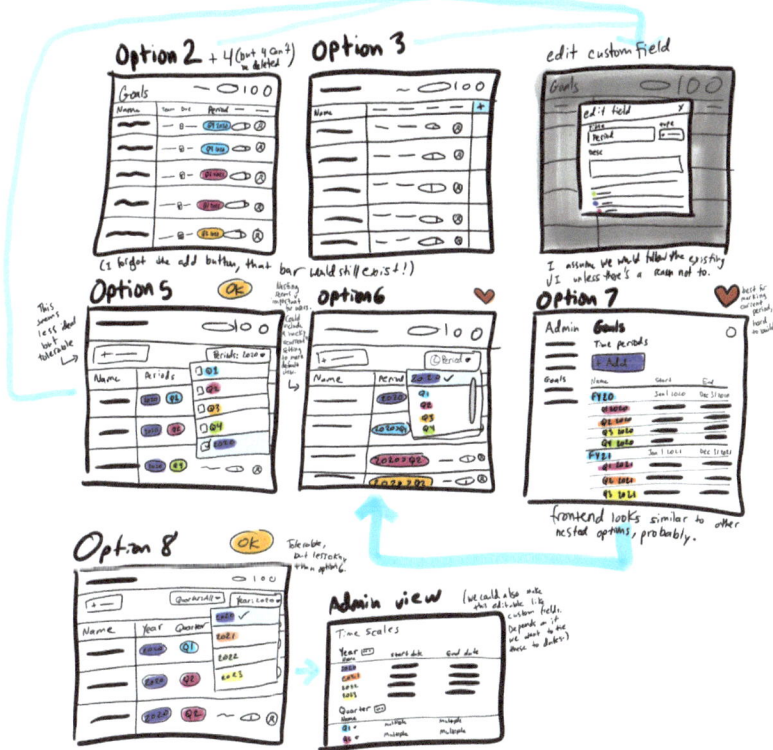

FIGURE 5.3
Hand-drawn sketches can show cross-functional partners how their decisions will affect the customer experience without taking up lots of design time.

All of these methods help teams recognize the state of the work. This enables teams to make directional decisions while also preserving design energy. It ultimately also ensures low-fidelity thinking won't get into the hands of customers.

Silos: Design in Public

Cross-functional partners and collaborators don't know what they can't see. Perfectionism discourages designers from sharing work until they feel it is as thoroughly considered as possible. This approach centers design work around production and removes the chance for teammates to provide early directional feedback. Silos increase the likelihood that a designer will move out of lock-step with their team, which often results in additional iteration cycles and scope cuts.

Instead, you should share work early and often so people can see how designs come together over time. Vulnerability is scary, but it has many benefits. Designers who show the messy process behind their solutions create more visibility into the complexity of design work. The more a designer shares with their team, the more the team can understand the decisions behind the final product.

CATT'S CORNER
Balancing Progress with Polish

Projects usually have time constraints, so staff designers must prioritize where they invest their energy. In Chapter 3, "Wrangle Your Time and Capacity," I talked about the value of collaborating with cross-functional peers to balance your workload. Designing in public lets the team see the results of your prioritization in live time.

At the outset of a project, I work with my product manager and engineer to make sure that we all agree with what parts of the user experience deserve more design investment. For example, when I'm creating a new feature that has a lot of complexity, I ask for a few extra cycles of design polish, so I can perfect the first-time experience. If we get the onboarding right, users will hopefully be able to repeat the process with a full understanding of how the functionality works.

It's also important to align with your team on which parts of the experience don't require as much effort. When working on products with robust design systems, for example, I tend to lean on pre-existing UI and industry standard best practices unless the user problem calls for a new solution. This is especially helpful when you're building on top of the work of other product areas. I unblock the team to start scaffolding certain parts of the experience by leaning into cross-product consistency, and then I spend most of my creative energy on choosing and validating the treatment for net-new concepts.

As I progress on a project, I follow through on the focus areas we agreed on at the beginning of the project. When I share progress updates, I remind the team of our agreements and point to the stage I'm at in the process. I also explain what's coming next so they're always aware of how I plan to spend my time. If necessary, we renegotiate the project plan based on what we learn as our ideas firm up. ∎

The Value of a Vision

Designers become strategic operators when they take hold of every opportunity to transform the product development process from an interpersonal battle to a collaborative effort. Rather than fighting against individuals in other roles, they work with cross-functional partners to combat user problems. In fact, teammates may start to ask proactively for more strategic support when they begin to recognize the value of design. One of the most common strategic artifacts that teammates request from staff designers is a vision.

What Is a Product Vision?

One day during the March 2025 cohort of my staff designer course, a senior designer asked me: "What is a vision?" Over eighty students had participated in prior cohorts. Not one had asked this wonderful question yet.

Visioning is the process of illustrating how a concept will play out in the near or far future. A vision can be created for entire products, product areas (as shown in Figure 5.4), or even small features. Staff designers create visions to help their product and engineering partners align on strategic priorities with senior leaderships.

A designer can make a vision that shows how their team's feature might be used by other product areas. Or they can make a vision that shows how their team's product area might evolve in two to three years. Sometimes, staff designers even facilitate the creation of company-wide visions that show how the entire product could evolve. All of these exercises illustrate ways the company can move forward and provide potential roadmap items for product teams.

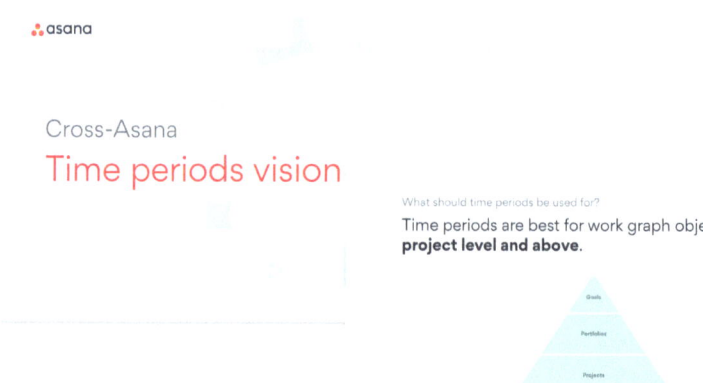

FIGURE 5.4
A vision presents strategic direction and creates clarity across teams.

> **TIP** **VISIONS ARE COMPANY-DEPENDENT**
>
> Every company is unique. The expectations of a vision are also unique to each company. Individual designers must align with their team to create a shared understanding of the value that product visions might offer, along with the formats that work best internally.

When to Propose a Vision

The goal of a vision is to illustrate the future of a customer experience. Yoko Sakao Ohama, Staff Designer at a major eCommerce company, believes visions help capture "held knowledge that a team has not been able to take action on." A vision is a powerful way to direct an organization or product team when they feel directionless. This feeling is most palpable when morale is low and an organization's roadmap is driven by external forces, such as unfiltered customer feedback or the highest-paid person's opinion. In challenging predicaments where the way forward seems unclear, a product vision can generate momentum and kick-start positive movement.

Visions are also useful when teams plan to embark on new major project work but aren't sure what to focus on. A constructive product vision inspires the team and can convince even the most stubborn individuals to invest in major product upgrades. It can also give a team the confidence to prioritize upcoming work with a strong opinion, supplying the perspective to push back on external influences when necessary.

Jen Pearce, Staff Product Designer at Mercury, recommends considering the context of the organization before embarking on a visioning journey. "Do I have something that needs to be said, and do I have an audience that is willing or able to hear it?" Every vision will be shared with an audience—"Audiences have varying energy and interest in visions." Jen recommends that designers confirm they "have the social capital required to pay for the airtime and venue to share that vision." This will ensure that the vision resonates with its intended audience.

How to Make a Vision

Like most design work, the process to create a vision depends on the context in which it is created. The needs of the individuals who will consume the vision are unique. Each vision serves a unique purpose and exists to deliver value at a particular point in time.

Although every vision is special, there are several common steps necessary to ensure that a vision is of value to a product team:

- Define the scope.
- Outline the narrative.
- Add polish.
- Socialize the final product.

Define Your Scope

Vision projects are a serious time commitment. Therefore, the objective and scope should be clear before the work begins. You can create a vision for an entire organization, a specific product area, or a new concept that a product team would like to propose. A vision can focus on the near future or illustrate a faraway amount of time ranging from one to five years in the future. The final scope depends on the stage and needs of the organization itself.

Visions must be aspirational but not encourage delusional thinking. Anthony Restivo, Staff Product Designer at Jasper, advocates for pragmatism: "You can spend a lot of time building this really cool concept that no one will necessarily care about because it's not possible or feasible." However, he believes in the value of showing "the best experience we can build that would help people do something they can't do today."

Like most design artifacts, vision work will be a waste if it is not consistently paid attention to. An ideal objective for a vision highlights the situations in which this vision will be shared or referenced once it is ready to be socialized. For example, a vision might be presented at a company all-hands meeting, linked in internal documentation, or consulted during a product team's roadmap planning. Stellar visions align with the business's strategic priorities and contribute directly to a product's roadmap, which automatically ensures that they will be continuously referenced in the future.

The timeline for building a vision varies but should also be agreed upon up-front. The process can be impacted by organization size, focus area, the number of individuals contributing key information, and other inputs. Senior leaders often underestimate the work that goes into creating a vision, so expectation-setting is key to carving out space to build a credible vision. From start to finish, you can expect to spend at least two weeks on a low-fidelity vision, at least four weeks on a mid-fidelity vision, and six to eight weeks or more on a high-fidelity, large-scale vision.

Outline the Narrative

Storytelling is critical to a compelling product vision. In 2003, Richard E. Petty, Leandre R. Fabrigar, and Duane T. Wegener conducted a study[1] that revealed that using emotions to convey a message can often be more persuasive and effective than rational arguments. Emotions are powerful, and your vision must lean into them.

A successful vision narrates how customers will move from their current situation, which is painful in some way, to a happy future in which the organization's concepts create a positive impact for them. Designers can define this narrative by gathering inputs through the facilitation of various exercises, such as the workshop shown in Figure 5.5.

A narrative includes characters and a story. In a vision, the characters represent the key audiences that will be affected by the future experience. The story covers the characters' development arc from current day to happy future.

---

[1] Richard E. Petty, Leandre R. Fabrigar, and Duane T. Wegener, "Emotional Factors in Attitudes and Persuasion" in *Handbook of Affective Sciences* ed. R. J. Davidson, K. R. Scherer, and H. H. Goldsmith (Oxford University Press., 2003), pp. 752–772.

FIGURE 5.5
A collaborator writes on a sticky note during one of several workshops I facilitated as part of a visioning exercise.

Every company has a series of individuals they believe to be their target audience. These individuals either directly or indirectly make decisions to acquire and use the products an organization builds. A vision may cover an organization's key target audience or focus on a subset of individuals within the user base who exhibit certain behaviors. For example, Maggie Yue, Staff Product Designer at Cisco, once created a vision to show how a product needed to support the admins that made purchasing decisions in addition to their most common users who didn't pay the bills. The characters chosen could highlight business opportunities and should therefore match the intentions of a product team, expressing who they will and will not accommodate in their illustrated future.

The narrative can be scaffolded once the characters are confirmed. At this point, the team can outline the major plot points they want to present within the vision. Information covered might include:

- Current customer pain points
- Business opportunities
- Key themes the team plans to explore
- Common customer scenarios
- Ideal use cases
- Expected qualitative or quantitative outcome of the team's investment

The goal of this outline is to form a sturdy foundation. While an illustrated experience is the final output, most teams therefore document this information by using a series of bullet points or sticky notes. This ensures that no energy is wasted, as multiple drafts and iterations may be necessary before alignment is reached.

Add Polish

Like all design work, a vision builds up over time. Once the team agrees on a foundational outline, the illustration work can begin. The driving designer must determine the presentation format and then begin the work of illustrating each step in the narrative.

Visions are usually expressed as either a series of illustrations (shown in Figure 5.6) in a slideshow, a write-up, or a prototype. A series of illustrations usually works well in situations where the audience is less familiar with the characters and context must be spoon-fed to them. An interactive prototype might work better when the audience has more context about the customer base and therefore needs less supplementary information.

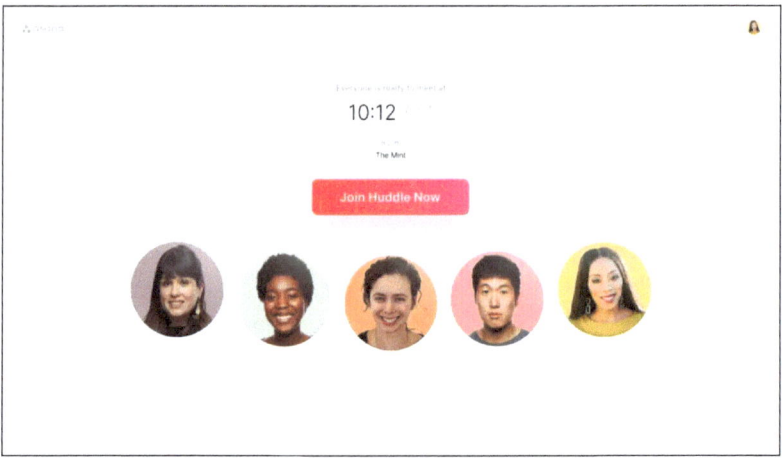

FIGURE 5.6
A visionary illustration from Asana, which shows a future in which a "virtual assistant will schedule the huddle and create an agenda for you."

The vision must include the perspectives of those who will be affected by it. Therefore, the driving designer must consult relevant individuals and teams within the contributor, observer, and approver categories for feedback. Feedback should be requested as soon as the initial draft of the illustrated vision has been completed. This is why it's important that the fidelity of the illustrations matches the fidelity of thinking—when iterations are required, changes will not be expensive to make.

With feedback and refinement, the vision will become clearer and more convincing. However, since a vision is speculative, there is endless potential for further adjustment. Like most efforts, perfect is the enemy of good. The timeline defined in the initial scope will provide the necessary bounds.

Evangelize

With polish complete, the vision will be ready for its audience. Many designers forget the value of this step; Kyle Turman, Product Designer at Anthropic with 15+ years of experience, said "The biggest mistake I've made—and other designers make—is thinking that once you're done, you're done." Distribution is the final step in the process, and it's key to ensuring the vision is a useful artifact.

There are many ways to share a vision. The best methods depend on the norms of the organization itself. For example, a live walk-through with Q&A might work well at a company that values synchronous collaboration; at a company that devalues meetings, an asynchronous video share-out linked in a message or company directory might fare better. Jason Huff, Director of Product Design at a major content publishing platform, recommends that teams set a goal of presenting at an all-hands team or other org-wide ritual: "You know the finish line is a meaningful share-out instead of having to create that moment for yourself."

Whatever actions are taken to share the work, the driving designer must ensure that the vision can be accessed easily by others long after they first share it. The vision should be linked to any relevant projects and shared in all relevant communication channels. This will make it possible for individuals to locate the vision without needing to directly contact the driving designer, which increases the likelihood the vision will live on beyond their employment.

In the best-case scenario, a vision will be broken down into roadmap work and assigned to teams or individuals for further exploration. However, many visions do not make it to this stage. If a vision is created and ultimately discarded due to lack of planned investment, that's okay—as long as it helped the team make that decision!

Debrief

Design has the unique ability to unlock strategy by increasing confidence and reducing risk. Even small actions, such as asking a curious question, can move a designer toward a leadership role on their team. Once they build trust with their collaborators, they can work with the team to co-create a vision for their product area or organization.

Designers sometimes give up their own agency without realizing it. A major difference between a senior and staff designer is the level of agency required to succeed in the role. Taking hold of your agency is a key component of becoming a leader.

It's our time. The seat is ready for us at the table. We need to show up, step up, and take it.

Activity

Define the scope of a visionary project for your product area or organization.

What is the objective of the vision you'd like to create?

Why would your team benefit from this vision?

Outline the individuals who must participate in and be aware of the vision creation process.

Individual	Job Title	Team	Role (Contributor, Approver, Observer)

Outline the potential timeline for the creation of your vision.

Stage	Milestone	Individuals Involved	Time Required (in weeks)
Outline the narrative	Choose characters		
Outline the narrative	Scaffold the story		
Add polish	First draft of full walk-through		
Add polish	Second draft of full walk-through		
Add polish	Final draft of full walk-through		
Evangelize			

Speak with 1–3 members of the team about the idea and get their feedback. If the idea seems worth pursuing, you can use techniques from Chapter 6, "Build Influence Without Authority," to get buy-in regarding the time investment.

CHAPTER 6

Build Influence Without Authority

How Influence Works	122
Pick Your Battles	126
Gather Your Research	130
Build Your Case	132
Share Your Case	136
How to Handle Disappointment	139
Debrief	141
Activity	141

Influencing the company at scale is a major challenge for staff designers because they have zero direct reports. Technically, no one has to listen to what a staff designer says because they have no institutional power by default. As an individual contributor, your ability to influence comes solely from how well you can convey ideas to the right people. By learning how to turn your opinions into persuasive cases that resonate with others, you can convince your coworkers to do what is right for your customers and the business.

Influence is usually a factor that keeps senior designers from being promoted to the staff level. Design managers often say that they can't promote certain individuals because those designers don't have enough influence across the organization. Lack of influence kept me from reaching the next level, too. It can feel so frustrating to hear this feedback.

What makes influence challenging is its intangibility. You know it when you see it. But you don't know how it works until someone explains it to you.

How Influence Works

Influence is the act of affecting changes in the beliefs, decisions, or actions of others. Iyobosa Irabor, Founder and CEO of the Insightfully coaching practice, notes that influence is a practice that is built over time. "Influence requires attention to how you communicate ideas, frame problems, time your contributions, and how you build agility into your stakeholder management approach."

Influential designers have a "persuasion toolkit" that includes storytelling, strategic questioning, relationship-building, and skillful advocacy. Iyobosa believes that credibility, or accumulated trust, is also a critical component of transforming the organization through others. "When others believe that you understand their needs and share their goals, credibility gradually evolves into durable influence."

Since change is the expected outcome, you must first understand how much you can influence each coworker. My favorite framework related to influence is known as the Locus of Control (Figure 6.1). The Locus of Control illustrates the amount you can directly or indirectly impact a situation. This framework was developed by American psychologist Julian B. Rotter in 1954. Originally connected to personality psychology, this idea has since been applied to the business world.

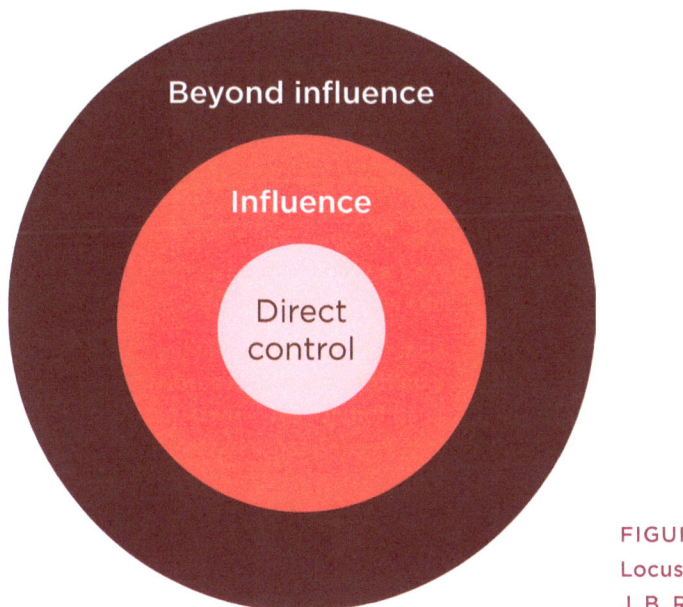

FIGURE 6.1
Locus of Control,
J. B. Rotter, 1966.

I learned about this framework after a project occurred where I failed to influence someone. In 2018, I was tasked with helping a product manager make a decision that impacted the user experience of our product area. I had eight years of professional design experience at that point, and I was deeply concerned that the product manager's proposed decision would result in a confusing experience for our customers. But the product manager wouldn't listen to me, no matter how much I tried.

After the project went poorly, I did a lot of reflection and stumbled on the Locus of Control. This framework helped me realize that we didn't have a stable, high-trust working relationship. The chances of me influencing this person were very low, if not impossible.

The Locus of Control has three layers. The innermost layer is you and every action you directly control. The second layer is people with whom you have a solid relationship and can therefore influence. Finally, the outermost layer is people that cannot be influenced. You must consider these layers when you consider the different problems you could solve, as they will impact your decision-making process.

IN THE REAL WORLD

MICAH BENNETT

Micah Bennett is a high-ranking designer at Figma with over 12 years of design experience. She and I met at Asana in 2020 when she joined as a Super-Senior Designer. We worked on the same team for nearly a year. I learned a lot about influence and storytelling from watching her work.

When we spoke, I asked her what influence meant to her. "Influence is when your perspective on things is well-regarded. Maybe leadership wants to put you on an initiative to solve a really complex problem because they have confidence in you. Or when you present an idea, you can see that they're actually paying more attention and listening." She believes it's the feeling that people trust your leadership and are open to your ideas.

To Micah, influence means nothing without seeing an effect on what gets built. "It's what's on the roadmap, what's being shipped, and the decisions being made for the product as a team." Micah is willing to flex, as necessary, to solve issues that some might find impossible. "I'm not a designer who just focuses on the design. I am influencing the product as a whole. How I work with others is critical, not just my single slice of craft." This mindset was clear even during our time together at Asana, and she grew to become a Director of Product Design within two years of joining the company.

Micah believes that recognition is also a signal of effective influence. She knows she has successfully built influence "if people in the company know my name and know of me in some way. If my name came up in a meeting, or someone saw my work, that tells me I'm showing up in some degree of influence as a staff designer." When others proactively share your work and speak about your work on your behalf, it means you've contributed to a shift in the way they think.

Layer 1: Direct Control

As you are yourself, you can control your own actions with relative ease. Junior individual contributors spend most of their time focused on this kind of work. They directly expend energy making a small-scale impact rather than working through others.

This layer pertains to your own actions and no one else's. You can only do as much as you have time for. There's an immediate dopamine hit each time you finish a task that makes direct control so addicting.

Direct control is not efficient, and organizations usually reward senior professionals less for this kind of work as time passes. Additionally, there's a risk of burnout if you take on too much for your capacity. The most efficient way to make more impact is to scale yourself. Scale results from moving beyond direct control into building influence.

Layer 2: Influence

The middle layer contains all the people you can directly influence. These are folks with whom you've built a solid working relationship. They trust you, believe in your abilities, and value your perspective based on your track record.

As a staff designer, your goal is to create as many of these kinds of relationships as possible so you can build the most influence possible. The more people who populate this layer, the more impact you'll have at the company. By nurturing your relationships as we discussed in Chapter 4, "Nurture Your Relationships," you should begin to build these kinds of relationships.

While you are developing your relationship with key individuals, look for cues that might denote a shift into this layer. Do they share unfiltered thoughts with you about the organization, ask you for advice, praise the outcomes of your work, or let you work with a high amount of autonomy? These are examples of positive signals that they trust your perspective and therefore may be able to be influenced.

Problems within your realm of influence are ideal for you to solve if you want to shine as a staff designer. While there are issues you cannot solve without the input of others, you likely have the connections necessary to be successful. People who focus their energy on these kinds of problems are seen as leaders because they can effectively align stakeholders and make a larger impact.

Layer 3: Beyond Influence

Some individuals are so passionate about a particular subject that they cannot be influenced about it. Others will be stubborn about every topic, no matter how hard you try to build a relationship with them. These are people who exist in the outermost layer beyond your Locus of Control.

These people must be factored into your decision-making process. Meghan Logan, Staff Product Designer at Thrivent, approaches these people as if they were participants in user research. "I try to reframe the conversation around fears and hopes. If someone is particularly stubborn, asking what they're afraid of and what they're hoping to achieve can open up a more honest, human dialogue." Many of these people have a lot at stake, and they are passionate because they are experiencing a lot of pressure.

Meghan said that gatekeeping may come from a place of duress, not control—especially when the pushback comes from a PM. "They're responsible for hitting business goals or key metrics, and sometimes design hasn't clearly articulated how our solution supports that outcome. When someone's performance is tied to delivery or success criteria, they often try to manage risk by controlling the 'how.'" You must acknowledge their opinions as you plan what problems to tackle, along with how to solve them. By understanding their boundaries, you can work with (or around) these people to find success.

Pick Your Battles

An early career designer might try to resolve nearly every problem they find. That outlook leads individuals to spread their energy thin, ultimately leaving them with burnout. Staff designers are more decisive about which problems they take on.

Staff designers need to show their value by delivering high levels of impact at scale. Consciously deciding what problems to solve can help you optimize for impact in your role, crystallizing your perception as a leader. It can also help you preserve social capital.

Social capital is the amount of trust and respect others have for your perspective. You can also view it as the strength of your professional relationships. Like personal relationships, you must invest energy into ensuring that your relationships with coworkers are as healthy as possible. And you must respect their boundaries, because not doing so will have negative consequences, ultimately subtracting your social capital.

There are so many problems, yet so little time. Criteria of your choice can help you filter through the noise and choose what problems to rally against. Lots of criteria can assist you. When I put out a call on social media about this issue, folks responded with several criteria to consider:

- Alison Gretz checks whether the problem "will be appreciated and known by stakeholders I need to influence."
- Asia Hoe finds out what kinds of "pain points [solutions will] introduce, and for whom."
- Yan Ling Tan asks, "Who in my team is impacted by this problem?"

My personal favorite three criteria are a bit different. I believe designers should look at business priorities, context overlaps, and ownership areas. These three criteria help you confirm whether your contributions will be impactful, you're a match for the work required, and your work will benefit your career long-term. The criteria Alison, Asia, and Yan mentioned can augment these three criteria and are valuable additional considerations for your process.

CATT'S CORNER
Influencing Org Design Challenges

Some of the challenges you face will be organizational, not customer-facing. Staff designers are not people managers, and therefore may not be viewed as the most credible source for org design recommendations. In situations where you identify an org design problem, you are welcome to propose a solution—but there's no guarantee that your approvers will follow through unless they feel the pain you're feeling.

In one of my first roles as a staff designer, a large software start-up hired me to drive the design of a new product area. This company had no hierarchy for individual contributors but featured several layers of management. Individual contributors weren't allowed to be area leads.

Despite the high priority of my work, I was three layers away from the head of design. This meant I had to present my work to three different managers before it reached the head of design. In some situations, I was able to negotiate and skip a level or two. But every layer wanted to be informed. At the worst point, I spent multiple weeks preparing review artifacts for each management layer because they all operated at different altitudes.

This was obviously unsustainable if the team wanted to ship work at the pace that was demanded by our customers and the business itself. So I put together a proposal explaining the value of putting people like me into leadership roles for entire product areas. I shared it with each management layer, explaining how the org structure affected my team's agility.

No matter how much I tried, the situation didn't change. I wasn't the person who could solve this problem, and senior leadership wasn't frustrated enough at the time to make the change for themselves. This issue wasn't resolved until a few years after my departure, and there are now multiple IC area leads. I'm happy for those who benefited from the change! ■

Business Priorities

Designers are deeply empathetic individuals, and that means they care about all kinds of problems. But all problems are not equally important to the business. You need to prioritize problems that meet the current needs of the business, as these will be more valued contributions and therefore more valuable to you as you build a track record of high-impact work.

If a company is in growth mode, problems that result in growth will be more rewarding to solve. If the company is focused on retaining employees due to a high rate of departures, contributions to organizational design topics, such as career ladder improvements, might be more valued by design leadership. And finally, if the company is facing a financial emergency and stabilizing the business is a top priority, that career ladder might be put on hold in favor of strategic work.

Of course, you might find a problem that *should* be prioritized by the business but isn't yet. If you have a large sphere of influence and can connect that problem to work that has already been prioritized, you might be able to get support to explore the problem area. We'll dig into this later in the chapter when we cover building a case. But be warned: Projects that aren't business-critical are less likely to get large-scale support.

Context Overlaps

There are lots of problems you can spend time on. The reality is that you cannot solve many of them, no matter how hard you try. Instead, try to find problems that match your unique skill set and knowledge, as these are more likely to be a worthwhile investment.

Being the right person to solve a particular problem is key to building a track record of expertise in a certain area. Senior leaders will be more likely to give you the green light if you are the right person to solve the problem. Focus on problems that are within your domain, especially if you have relevant historical context. For example, if you have a history of working on monetization at other companies, you might be the right fit for a growth design project.

As a hands-on designer, some pains are not yours to heal. Aim your energy at issues that match your experience and delegate ones that don't. You are welcome to highlight extraneous problems that impact you as an individual. Share these issues with your manager or other people who are the right fit. Let the responsible parties decide how to handle them.

Ownership Areas

Designers often get themselves into situations where they try to fix problems that don't match their current role or their future career ambitions. Even if you find certain issues infuriating, you must be strategic about the ones you take on. One example of a way to think about this is to ask, "Would I enjoy being the owner of this solution in two years?" If the answer is no, you might want to let someone else handle the problem. If it's that big of a deal, it will eventually become annoying enough for others to notice.

As empathetic and detail-oriented people, designers often predict and fix problems that other people don't mind tolerating. When you fix these problems, other people refrain from stepping up because they know you will come to the rescue. This phenomenon is commonly referred to as *glue work*, a term popularized by Tanya Reilly, principal engineer and author of *The Staff Engineer's Path*, through a technology conference presentation and blog post she shared in 2017.

When you do glue work, you subtract your own capacity. This results in less time for the kind of work that benefits your career. For example, you might notice that your company could improve the way they onboard new designers to the team. Fixing this problem might seem like a good idea in abstract, but this kind of work could result in external pressure to shift into design operations or management.

Make sure that you build the kind of track record you want to have. If you want to continue being an individual contributor, then narrow your focus. Only take on problems that demonstrate your ability to deliver high-quality, thoughtful experiences. As challenging as it might seem at first, it can also be liberating to let others own problems that are rightfully theirs.

Gather Your Research

Once you've identified a problem that meets your initial criteria, you must spend time learning as much as possible about it. Thoroughly researching the problem will help you assess the likelihood of your ability to influence the outcome of potential solutions. This process will also generate enough sources to craft a compelling pitch.

Several types of insights you can gather in relation to a problem include organizational, qualitative, and quantitative research. Each kind of research is a building block that contributes to the overall story you can tell. The more you compile, the more detail you will be able to add.

Make sure that you consider the key questions you need to answer to confidently state your case. Examples of questions might include:

- What did we try in the past, and why did or didn't it work?
- How many people does this problem affect?
- What frustrates users about the current experience?
- What technical challenges might limit our ability to change the experience?

Proactively focusing on key questions will help you avoid over-researching the problem. Once you have enough of each kind of research, move forward. You can always incorporate additional research as you identify gaps.

Organizational Insights

Internal research helps you understand the company-facing causes and impacts of a particular issue you've recognized. This research also gives you an understanding of the internal blockers you will need to overcome to make the problem go away. Blockers can be any number of things, including but not limited to people, teams, divisions, and technical systems.

Gather this kind of information by connecting with other relevant individuals who might have some historical context. Short, exploratory conversations of 30 minutes or less can uncover lots of relevant insights to dig further into. Perhaps there are senior leaders, product managers, engineers, or other designers who have previous experience with the problem. What challenges did they face? What did or didn't they try? Do they believe the problem can be solved in the future? These conversations can help you chart a path forward, all while identifying potential supporters and skeptics.

Workshops can be a great complement to your exploratory conversations. A one-hour session with people in similar roles or with a similar domain of expertise can uncover a lot of detail in little time. For example, someone might gather all the designers who work in a similar area and explore the problem together. This illuminates the path forward while enabling the cross-pollination of ideas across teams. It can also generate the momentum necessary to inspire others to advocate for the prioritization of this issue—and the implementation of solutions—on your behalf.

If you work at a large, established company, you can also dig into internal documentation to learn more about the problem. Systems where your company hosts project briefs, product specifications, design files, and other forms of information might include facts that can aid your case. Maybe old research or data from past experiments are relevant to the problem you intend to solve.

The scale of your company is directly correlated to the likelihood of you being able to access relevant past written documentation. Small start-ups tend not to document past work very well because individuals are moving quickly. This is especially true for individuals at companies with under 100 employees. If you work at a new or small company, expect to rely on verbal history. Meet with individuals who might have context and ask them to share as much as they can about the subject.

Qualitative Insights

Designers are connected to the pains of their users and customers. Just like all other forms of design work, the lived experience of your users and customers should be factored into your advocacy efforts. Speak with people who use (or could use) your product to uncover the emotional reasons the problem must be solved.

Building a research-backed perspective doesn't require you to plan and schedule user research. Since you have yet to receive the green light to focus on this issue, gather qualitative research using lightweight methods. Your company's customer support team can provide you with a summary of customer feedback. You can search through feedback survey submissions, customer support forums, and even social media websites to find relevant customer complaints.

As you get signals from customer insights, you will start to form a narrative. User complaints about harassment can support an investment in trust and safety tools. Customer feedback about the lack of useful data can support the campaign to redesign a product's dashboard. These kinds of insights are super valuable and help you identify the human impact that will result from an investment in solving the problem.

Quantitative Insights

Statistics illustrate both problems and opportunities. Companies hate to leave money on the table. You can get support for your initiative by showing how that is happening today.

Numerical data gives you confidence that solving this problem will have the meaningful impact on the business you hope for. *Scale* is the key to getting support for your initiative. *Metrics*, such as conversion rates and annual recurring revenue, pair well with qualitative insights to strengthen your narrative. A project that impacts a large paying user group will be prioritized more heavily than a small group of free users.

The more detailed analytics you can access, the clearer your story gets. For example, a data-driven customer journey chart can illustrate areas where users drop off during a flow. This can underscore the case for redesigning a form or checkout experience.

Smaller organizations often don't have great data sources. If you work at a start-up, you will likely have access to very little data (if any at all). If your engineering team has yet to implement an analytics tool, put it on their radar. This won't just be key to further clarity for your work—it will also help the engineering team perform better. Lean on your design manager, product manager, data scientist, and tech lead; they're responsible for implementing an analytics tool. While you wait for a richer data pipeline, rely on your intuition.

Build Your Case

With your research in place, you should have enough information to decide whether you should drop the issue or move forward with a proposal. If you have conviction that the problem is truly worth solving, it's time to build a compelling case. As shown in Figure 6.2, you can build the outline for your case using a formula comprised of three parts:

- **Observation:** An explanation of the problem
- **Proposal:** Potential solutions
- **Outcome:** The expected result once the problem is solved

OBSERVATION		PROPOSAL		OUTCOME
Our upcoming release is too big.	+	**Incrementally add updates over time.**	=	**Higher chance of launch success.**
We can't measure the impact of each change.		Here's a write-up of a draft rollout plan.		We can tell which changes are more or less impactful.

FIGURE 6.2
An example showing how you can use the formula to pitch your idea to a product manager.

Each part of the formula is equal in importance. Together, they are three building blocks of a potent pitch. If you provide an observation and proposal without an outcome, senior leaders won't be able to predict the impact of your work. And if you put forth a proposal and outcome without an observation, the reason for doing the work will be unclear.

As a senior leader, you must make it easy for other leaders to enthusiastically agree to your plans. The problem should be so clear that the solution is a no-brainer to accept. If a key piece of decision-making data is missing, they cannot provide consent. When you build confidence in your ideas, you receive trust from your collaborators and can ultimately better influence the direction of the organization.

Observation

The observation is an executive summary of the challenges you compiled during your research phase. It should clearly state the internal and external pains that impact both your organization and its customers. Its goal is to prime the audience to receive your proposal.

Use as much detail as necessary to show your audience that the pain is real. Customer quotes and research video recordings can be very convincing. Screenshots of the current in-product experience can help for situations where a product is buggy or frustrating to use.

Along with the user experience, also use this section to outline the business opportunity. Add any relevant data that might help different audiences understand the scale of the problem. Product leaders might want to know how this problem impacts growth metrics, such as monthly active users, whereas a director of product design might care more about the net promoter score for the area you're discussing. Including content for different audiences can help you angle the problem to those with unique needs while using the same artifact.

Finally, acknowledge any past work that has been done in the problem space before. This is especially important for navigating the opinions of skeptics who have been employed at the company for a long time. Explain why past attempts failed and how your plan will result in a better outcome.

Proposal

Your audience will be ready to hear your proposed solution once the problem has been clearly explained. At this stage, you don't need to have all the implementation details for a fleshed-out solution. Your proposal can be as simple as a plan to define a direction. The goal is to get your project prioritized and on the team's roadmap.

Junior designers often observe problems but don't propose solutions to them. They find it daunting when leaders ask them to bring solutions, not just problems. This is because junior designers speak out about a lot of problems, regardless of their size. When you are more selective with your energy, you become more efficient and will find this experience less frustrating.

The truth is that senior leaders are busy, just like you. Coming to them with solutions will help them help you. It will also help them see you as the leader you are, and it will help you feel less like you're doing their job for them.

Since the work you propose will directly benefit you, you'll be more excited to take it on. Once you present the proposal to leadership, they will likely be intrigued. But they need one more detail before they can proceed: the outcome.

> **TIP** SHOW A FEW OPTIONS
>
> Proposing a single solution might intimidate approvers; they may not feel informed enough to commit. Instead, come to the conversation with a handful of options so they can choose the

one that works best for the business. Proactively communicating multiple alternatives to resolve an issue will lead to a smoother discussion. Your thorough approach will allow you to steer clear of questions about why you didn't go in another direction.

When you list out a few options that would resolve the problem, be sure to include pros and cons of each option. Highlight which solution you believe will result in the best outcomes for the business and your customers. This empowers approvers to quickly understand your thought process and either consent to your recommendation or provide informed feedback.

Outcome

Wrap up your proposal with the expected outcome of the work to get their optimistic buy-in. Projects can have many outcomes. Choose an outcome that is positive for each audience member you must convince. Ideally, the result of this investment impacts the overall business in addition to the customer experience.

Different senior leaders and collaborators care about different outcomes. A product manager cares about risk management, so you might explain how your proposal increases the likelihood of a successful launch with higher metrics. An engineer cares about code quality, technical stability, and simplicity so you can tell them about the ways your solution will decrease the release of bugs in your product. A design director cares about user experience quality so you may communicate the expected outcome in terms of an improved user experience or brand perception.

Business impact is key to getting your work prioritized. It can be additive, subtractive, or qualitative. Your goal is to estimate which one will give you the most impact possible.

Additive Impact

Positive metrics that are meaningful to the business, such as revenue, average revenue per user, and conversion rate, are all examples of *additive impact*. These are numbers the company wants to increase. A proposal that features these kinds of outcomes is most compelling because it results in an increase to the company's bottom line. This result can be used to show growth, which impacts the financial value of the company.

Subtractive Impact

Outcomes that decrease a negative metric are known as *subtractive*. Instead of contributing positive value, they remove detractors, such as the number of customer support tickets or annual spend. These kinds of numbers are a bit less compelling to the business as they show a decrease in expenses rather than an increase in revenue. This might be useful in situations where your company needs to show fiscal responsibility.

Qualitative Impact

The least valued kind of outcome is *qualitative*. This might include metrics such as Net Promoter Score (NPS) or internal user experience grades and measurements. These numbers are generally less valued by companies because they are not usually directly tied to revenue.

In my experience, qualitative impact serves best as a secondary or tertiary metric used to build alignment with design and research leaders. However, a CEO with a design background might value these measurements since they connect to experience quality and brand perception. Even if your company doesn't value them yet, highlight them in conversations with senior leaders to increase their visibility over time.

Share Your Case

Your completed outline contains all the information necessary to communicate your pitch to different audiences. You'll now need to craft artifacts in formats that will resonate most with your audience. Every leader is different, and strengthening your relationships with them will help you understand which formats to use along with which delivery methods they prefer.

Presentation

There are lots of ways you can present your work. Only some of them will successfully convince senior leaders your idea is worth the investment. Luckily, seasoned designers already have lots of experience crafting documentation in various formats. This will come in handy!

Foundation

The foundation of your presentation is the *format*, which determines how compelling your case will be to your audience. Textual presentation formats such as *write-ups* work better with more analytical individuals such as product managers, data scientists, and engineers. Visual formats such as *slideshows* take more effort but are more effective in situations where your audience needs to be spoon-fed information.

Senior leaders who prefer not to read nitty-gritty documentation will generally prefer slideshows. A slideshow can only contain a small amount of information on each page. This requires you to simplify the presented concepts as much as possible, which will help your busy audience wrap their minds around your idea.

Efficiency is key to choosing a winning foundation. An amazing slideshow that doesn't meet the audience where they are is ultimately a waste of time. When a write-up can serve the purpose, conserve the energy you would have spent making slides and use it on other work.

Illustrations

Components that demonstrate more details of your idea within a presentation can help your audience understand your intended outcome. The type of illustration you include depends on the type of problem you intend to solve, the fidelity of your thinking, and the needs of your audience. Examples of illustrations include sketches, wireframes, mock-ups, prototypes, and annotated designs.

Sketches illustrate an idea while setting the expectation that the work is not yet ready to be built. These work well when you are communicating with individuals who can extrapolate and imagine the intended future by seeing a low-fidelity drawing. Unfortunately, some senior leaders cannot do this. You'll have to spend more time creating higher-fidelity artifacts that resonate with them.

Wireframes are higher fidelity than sketches but are still clearly conceptual artifacts. These take a relatively low amount of energy and can show the gist of an idea without setting expectations regarding the final product. Like sketches, wireframes work well with highly imaginative senior leaders. While they are more illustrative, they still might perform poorly with those who cannot extrapolate an outcome without seeing high-fidelity examples.

Mock-ups are high-fidelity designs that align with the product's visual styles. They look ready-to-ship, which can help senior leaders see and understand the end state. But because they look finished, they generally create the expectation that the design has already been done. If you need to create mock-ups, make sure to apply a color overlay so teams know the designs can't be shipped yet.

Prototypes, also sometimes referred to as *vision types*, are one of the highest-fidelity illustrations of an end state that you can include in a proposal. Senior leaders love prototypes because they demonstrate the future with a high level of interactivity. Prototypes can inspire and motivate teams to prioritize work that results in a better user experience. Because they are usually more time-consuming to build, you should only use them in situations where interaction design details would benefit your case.

Annotated designs are another example of the highest-fidelity illustrations you can create, and they are most useful when the solution is clear. These are perfect for highly technical audiences, such as your engineering team. Avoid showing these to senior leaders, as they are too much in the weeds.

Artifacts

Diagrams, *personas*, and *journey maps* are a few examples of illustrative artifacts that can help explain the qualitative customer experience for your audience. If you decide to create them, reuse them as many times as you can. Designers often create these kinds of artifacts because they look cool. But if you only use an artifact once, it's not worth the time investment.

Delivery

After you prepare your presentation, it's time to deliver it to your audience. Delivery can be *synchronous* or *asynchronous*. If you have the time to deliver a synchronous presentation, this will be most impactful because you can answer questions live. Ideally, each synchronous conversation has fewer than 10 participants so you can have a candid discussion. If you need to have multiple conversations, this route might be untenable.

Asynchronous presentations can save you time by enabling you to present the same information to different audiences in less time. The best asynchronous presentation format is a video recording shared in a team space. This enables viewers to engage with the content and ask questions.

You can combine both types of delivery based on the required scale of your message. For example, perhaps you present synchronously to four senior leaders during half-hour meetings and then send a video recording to the broader team. With this kind of combination, you craft a white-glove experience for the decision-makers and ensure that observers are also properly informed.

How to Handle Disappointment

You'll have more influence in no time if everything goes according to plan. Unfortunately, there will still be times where you fail to influence others. The company will undoubtedly decide *not* to prioritize your ideas sometimes.

I've experienced this kind of rejection many times. It can hurt and feel disappointing, especially if you strongly believed in an idea. *When*—not *if*—your ideas get rejected, remember that this is a completely normal part of the process.

You can handle disappointment in two key ways. You can appeal the decision if you have the conviction, energy, and social capital to invest further. Or you can decide to move on. Both actions are equally valid.

Appeal

After receiving a rejection, you have the option to push back. There's no harm in asking for more detail regarding why it's the wrong time to prioritize your idea. You can use the feedback you receive to make a great follow-up case.

Appeal at least once to test the waters, especially if you are a people-pleaser. Persistence can be rewarding, especially if you have the social capital and can connect with those who could persuade senior leaders who are blocking your idea. Perhaps the feedback you receive will unlock the next steps you can take to get the work prioritized.

It's possible that the key blockers you tried to influence are outside your Locus of Control. If you are close to someone they trust, that trusted person might be able to assist you. I've had luck with offhandedly mentioning my challenges to other leaders who supported my proposal; they went and applied pressure on the stalwart resisters.

Tread lightly when you make your appeal. There might be formidable reasons behind the rejection that you are not privy to. If you push back with too much force, you will also be labeled as a negative boundary pusher. If someone above your pay grade declines to move forward with your idea, no matter what you say, it's absolutely okay to put it on a backlog for later and refocus your energy.

Move On

A very healthy response to rejection is to move on. The idea can be saved for the future, and you can bring it up when the right moment occurs. Or perhaps the idea isn't a good fit for the company's long-term direction, and you save it for your next employer. Whatever the case, you must fully commit to refocusing your energy if you decide to move on.

When you agree to move on, document and share the rationale so it can be referenced later. For example, if a designer wrote a proposal for a feature and it got declined because the team doesn't have capacity, they could add a note to the top of the write-up and cite the team's capacity as the reason the project was deprioritized. This is commonly known as *keeping receipts*.

Keeping receipts can be a valuable practice, especially if your proposal gets rejected. Someone else may pick up the problem in the future and reference your work. Or perhaps the information will be useful once you have the buy-in to pick it up in the future. Documentation can also be useful in case an approver tries to pass the blame onto you for their own past actions.

Moving on can be a very healthy response because it acknowledges the reality that you cannot control the outcome of the situation. It also ensures that you won't burn out at work. One of the biggest causes of burnout is feeling a level of ownership over your work that is out of sync with your job description.

Despite the way it may seem in the moment, you are not solely responsible for the outcome of your work—especially this kind of advocacy. Someone else said no. That person was the decision-maker, and you could never control them no matter how hard you tried. If you have unyielding belief in the idea, archive and save it for another day.

Debrief

Influence is a critical tool in a staff designer's toolkit. Many designers struggle to gain influence, and they spend social capital on irrelevant problems. Designers must ensure that they invest their energy appropriately by choosing business-critical problems that match their unique context and skill set.

By using information gathered from the process of building solid relationships, a designer can compose a compelling argument for their ideas. Insights from qualitative and quantitative sources can make an argument even more persuasive. And when the right presentation format is used, an idea can become so compelling that leadership has no choice but to get on board.

While some proposals will be a slam dunk, others won't be as well-received. Detractors are often willing to share feedback that can help a designer improve. But sometimes it's best to move on and preserve the working relationship. Maybe the idea will be worth investing in after time passes!

Activity

Choose an issue you want to influence.

What is the issue?

Why does this issue matter to you?

In 100 words or less, write a high-level description of the solution you think would resolve the issue.

Use the observation-proposal-outcome format to explain your solution to a few potential stakeholders and collaborators.

Collaborator to influence	Observation	Proposal	Outcome

Choose one artifact with which to try influencing each of the above individuals.

Collaborator	What artifact(s) would best influence this person and why?

Before creating the artifacts, connect with each individual in a live conversation. Discuss your observation, proposal, and outcome. Float the idea of creating the artifact(s) you came up with and then spend some time making a first draft if they are amenable to it.

CHAPTER 7

Scale Your Impact

Know When to Delegate	146
Find Work to Delegate	147
Choose How to Delegate	151
Set the Bar	156
Debrief	163
Activity	164

Staff designers usually don't work by themselves, and they can't make a major impact without the contributions of others. Delegation is an important part of the work, as we all have the same number of hours in the day. If you want to grow your areas of impact, you need to let go of lower-level work and instead focus on the creation of positive design outcomes at scale.

Unlike design managers, staff designers continue to be hands-on design leaders. However, they are selective about how they spend their time. When you optimize for positive effect, it's not about the number of pixels you push but rather the size of the change you can create. Delegation with intention will help you ensure that the quality of your work stays high once it's out of your control.

Know When to Delegate

Finding yourself stretched thin lately? Delegation is a key piece of scaling yourself as a staff designer. Consider what you might be able to take off your plate and share with others.

Being overwhelmed by your workload isn't a moral failing. Every designer reaches a point where they have more work than time. It's natural to feel like the problem is you. But no individual can take on every single piece of work—that's why you need to prioritize projects and maintain backlogs!

Like everyone else, you are human. Humans have limits. Use this fact as an opportunity.

When you choose some work to share with others, you give them a gift. They will learn and grow as designers. When you deprive them of this gift by taking all the work on yourself, you both bury yourself under an unsustainable workload and isolate yourself as the sole source of knowledge. Delegation can be an act of self-compassion and an investment in the future of others.

Self-reflection is required before you can externalize your work. The internal work to undo individualistic heroism can take time. You'll need to learn several things:

- **Delegation means giving up control.** The shift from direct control to influence is scary, especially the first few times you delegate work. If you don't learn to let go, you will become a micromanager and likely stunt your own career by reducing the amount of impact you can make.

- **Delegation requires trust in others.** Just as you don't want to be micromanaged, the individuals who will execute the work you assign them also yearn for autonomy. Build a solid working relationship with them so you can trust them to deliver in your stead.
- **Delegation enables you to open other doors.** When you give an opportunity to someone else, you create the bandwidth necessary to tackle larger scopes of work yourself. Building an abundance mindset will help you orient yourself toward what matters most to you, your team, and the business.

The delegation process can be intimidating at first. You must decide what work to delegate and to whom, and each of these decisions affects the likelihood of success. Additionally, you'll need to be present to steer the work to ensure that it meets the company's quality bar. But once you get comfortable with the change, you'll start to reap many benefits!

Find Work to Delegate

Delegation is a process that involves giving the right work to the right people. As shown in Figure 7.1, every project is not worthy of delegation. If you give an overly complex project to a junior designer, that is setting them up to fail. And if you delegate a simple project to a senior designer, they will become demoralized. You also want to work on interesting projects yourself so you can continue growing as a designer in your own right.

FIGURE 7.1
Focus your energy on the highest-impact, most complex work and either delegate or park the rest.

IN THE REAL WORLD

SUNNIE SANG

Sunnie Sang is a Staff Designer at a major social media company. Over seven years, she grew from a mid-level designer to senior and then from senior to staff. While she recently transitioned from a staff-level position to a manager role, we spoke before she made the switch.

As a staff designer, Sunnie worked across multiple designers on a number of occasions to deliver end-to-end customer experiences. "There will be projects that have two to three designers. Each of them own a specific area, but it ladders up to a larger initiative." Sunnie's role was to influence strategy and then steer the team in the right direction as they delivered design work.

Sunnie engaged with other designers on the team at a regular cadence to ensure they had enough support and guidance. "The team has weekly working sessions. But we don't wait to meet. We'll just chat about designs with each other." By encouraging the team to share proactively, even outside the weekly meeting, Sunnie helped create an environment of transparency and safety. This gave designers the ability to be vulnerable with each other and ultimately helped everyone get more timely feedback.

In these synchronous and asynchronous moments, Sunnie helped to connect the dots between designers. "I might have more insights into what everyone's working on, so I can connect our team with another team that I'm talking to." Sunnie's work helped collaborators—and other teams that needed to stay informed—move toward a mutually beneficial outcome.

With all these variables, how do you choose what to let go of and what to keep? If your work fits into the following categories, it may be worth delegating:

- Approachable
- Moderate impact
- Low conflict
- Low complexity

You won't always be able to get rid of work in these categories, but you should try as often as possible. Similar to how you decide which projects to advocate for, the goal is to spend most of your time working within your zone of genius. Delegation ensures that you're making the most impact as often as possible.

Approachable

Staff designers are expensive. Using their time on work that could be done by a less experienced designer without much context-sharing is especially expensive. Projects of this nature are great candidates for delegation, as they can be picked up easily by others.

Jen Pearce, Staff Product Designer at Mercury, considers if she is "uniquely qualified" to do the work. If she is not, she will consider delegating the project to someone else. She looks for individuals whose careers could benefit from that work and sends it their way—for example, handing a project to someone who could use " the opportunity to showcase progress in a growth area."

Projects that require a lot of insider knowledge are not easy to delegate unless the assignee has the relevant context. The hefty ramp-up effort is often not worth the time, because the individual you delegate to will often ping you with questions. If you would need to heavily administrate the project, you might as well just do the work yourself.

Moderate Impact

Certain parts of the product development process will yield more impact than others. Staff designers are often expected to focus their energy on visioning and direction-setting, which often unlocks multiple teams' worth of work. This is only possible if you hand off projects that have less of an impact.

Occasionally, you'll need to take one for the team and do more hands-on tasks, such as preparing design specifications for engineering. As often as possible, check if your current project work is the most impactful use of your time. If the answer is "no," it's time to refocus.

Low Conflict

Projects with little overlap or dependencies on your key focus areas are fantastic candidates for delegation. While you should have regular input regarding the direction of any work you delegate, you should avoid being a blocker. If the work depends on major design decisions you've yet to make, the project will hit a point at which it can't move forward without you.

Running two projects in parallel can be stressful for all parties; it's challenging to ensure that you always make important, unblocking decisions ahead of the other track of work. Whenever possible, delegate design work that relies on information that already exists. This will make your life easier, as you can prioritize your workload without concerns about the assignee's timeline.

In cases where you must delegate work with overlap, assign it to a seasoned designer who is confident in their ability to navigate ambiguity. Since the overlapping portions of the work will still be in progress, they'll need to work in lockstep with you. Schedule regular check-in meetings to ensure that both parties are sharing key information.

Never delegate work you'll depend on for your own design process. As the project lead, you should own the most important work with the highest number of dependencies. You'll otherwise end up undoing others' decisions if you disagree with them, and that can burn through trust. At the beginning of projects where you know you have strong opinions on a subject and suspect your capacity will be low, facilitate collaborative design time as the work kicks off. This will enable you to have impact on the thinking behind the work before it's too late.

Low Complexity

The projects you delegate shouldn't be overly complex, as you'll likely be giving them to people who are less experienced than you. When you hand over complicated projects, you decrease the amount of time you'll have for your own work. Steering individuals who take on complex projects can take many hours of live and asynchronous time per week. This is especially true for less senior designers, who generally require more assistance and feedback.

Small- and medium-sized projects with relatively simple user flows are best for other designers to take on. You'll be able to onboard them without much overhead. Since the work for projects of these scales is relatively predictable, you can then check in with the assignee at regular intervals without worry.

Choose which work to delegate based on who will receive the assignment. An individual's design manager can help you assess their capabilities and identify which projects will be just challenging enough for them. If the individual is newer to the company or you have less experience working with them, choose a less complex project since it's simpler to track and direct. You can let them know the goal is to calibrate your understanding of their skills for future collaboration.

Choose How to Delegate

You can delegate work in lots of ways. Your options depend on the scale of your company, how well-funded it is, and how well-resourced it is. The ideal form of delegation is an internal process that involves adding a person to your team long-term. There are several ways to do this—some of them involve major HR processes, while others don't. Less ideal methods of delegation feature short-term assistance, because they remove the opportunity to build shared historical context and work experience with the people you will continue to collaborate with long-term.

Delegation methods can be categorized as either *internal* or *external*. *Internal delegation* refers to temporary or long-term shifts in assignment for those who are already employed. *External delegation* involves either hiring a temporary contractor or a new, full-time employee.

> **TIP BRING SOLUTIONS TO YOUR MANAGER**
>
> A staff designer's manager usually makes the final call about any staffing decisions. But staff designers are also leaders, so they should lead with an opinion. Therefore, delegation requires a designer to influence their manager. Staff designers must consider the different modes of delegation, choose their preference, and then take a proposal to their manager to get buy-in. If the manager disagrees, they can work together to reprioritize the designer's workload.

Internal Delegation

Finding other designers at the company who can support your work is a great way to delegate work. It's more cost-effective to reassign an individual who is already on the payroll than to open a new position and spend hours on recruiting someone new to join the team. Three ways to make internal moves are partnerships, rotations, and re-orgs (see Figure 7.2).

FIGURE 7.2
The three types of internal delegation.

Partnership

The lightest-weight internal delegation method is a partnership. In a partnership, you collaborate with another team to get projects from your backlog prioritized on their roadmap. The partner team designs and builds your work on their schedule.

A superb benefit of partnerships is that you don't need to spend as much time on oversight. Your team receives support with the design, build, and rollout of a solution that both teams agree with. This usually results in a major win for both teams and breaks down silos at the same time, which can have positive long-term effects.

The downside of partnerships is that you are beholden to another team's timeline. They might slow down work or deprioritize your project completely if their goals change or more pertinent issues arise. Partnerships work best when you have a problem that overlaps with another team's remit. This ensures that you'll all be equally committed to the work.

Get buy-in from senior leadership before agreeing to a partnership. Senior leaders must be present to apply pressure whenever necessary. This ensures that the partner team will continuously prioritize your work.

Rotation

A designer from another team can join your team for up to three months to help with a specific project. This individual likely won't be reassigned to your team long-term and will therefore keep the same manager. However, this designer might have relevant experience or an interest in the subject matter you cover. A major benefit of rotations is the relatively short time to onboard, but the collaboration is temporary, and the individual cannot commit to any future support.

A major downside of rotations is that you must heavily manage the working relationship with this individual since they are essentially an internal consultant. You're also taking this person from another team, and that might ruffle some feathers unless the other team volunteered to share them with you. And when the rotation is complete, the assignee will return to their original team—meaning you might be understaffed later.

Reorganization

The most extreme version of an internal delegation is a re-org. This involves permanently taking a designer from another team and onboarding them to yours. Since this is an org design adjustment, re-orgs require staff designers to consult with design managers and influence the appropriate change.

If you have lots of work to delegate and another team is low on design projects, this solution will be worth the logistical overhead. The individual who joins your team will build up their own long-term context by taking on lots of projects. Eventually, they can lead work for more complex problem spaces.

Challenges might include managing the initial shock of the change. Since this is a long-term change, this individual will be shifted into your reporting chain and potentially report to your manager. They'll also need to attend all team rituals. As they ramp up, there will be drag associated with creating new norms as a changed team. Integrating this new individual within the team structure will require a concerted effort.

> **TIP** **RE-ORGS CAUSE CHURN**
>
> Re-orgs are time-consuming, expensive, and frequent—especially at large companies. In 2013, a McKinsey & Company survey of over 2,000 executives found that over 70% of reorganizations reportedly failed. Most organizational redesign efforts last less than 12 months. In my lived experience, this has also been the case.
>
> If you successfully orchestrate a re-org that results in a designer joining your team, please note that it will be disruptive to the designer, your team, and the designer's former team. That individual will need to onboard to the context of your team, and you may need to repair some broken alliances depending on the reception from the designer's previous team.
>
> After their onboarding is complete, cherish the time you have with your new teammate. Since re-orgs happen once a year or more at most organizations, you may not be working with them for long. Make sure that your management chain can see the positive impact of the change using communication techniques from Chapter 8, "Show Your Value."

External Delegation

Outside hires can be beneficial if a company has the budget or head count to bring on a new individual. This process involves adding a person to the company for either a short- or long-term period. The two kinds of external delegation are temporary and permanent hires.

Temporary

Short-term, freelance workers can help bulk up your team when your employer is unable to sponsor a permanent hire. This individual will have to sign a temporary contract for a specific scope of work in addition to a nondisclosure agreement. Their hiring will most likely need to be approved by senior leadership along with your company's legal and finance departments.

Unless you work with the same contractor every time, you'll have to onboard this individual with intention. They likely won't have any awareness of the company's goals or internal politics. You will need to give them insight into both your product area and the company's overall priorities.

Since this person is a temporary worker, you will be responsible for managing the working relationship. The contractor will not participate in most team rituals, and they may not be able to fend for themselves in most meetings with senior leadership unless they are very experienced. This gives you an opportunity to showcase your creative direction and leadership skills.

At the beginning of the project, create a brief that explains the scope of work in as much detail as possible. Outline the milestones at which you'll check in and clearly explain the definition of project completion. Since the individual is a consultant, they need the project to be very straightforward and the deliverables should be agreed upon up-front. If the scope ever needs to change, they may require a new contract and additional payments.

While contractors are helpful in a pinch, they can require a lot of overhead in the long-term. The contractor will roll off your team and be removed from the company, including all internal communications, at the end of the engagement. As part of their offboarding, they will hand over all design files they created along with any other relevant documentation they created during their time on the project. If you want to work with the same individual in the future, you will have to re-engage them, agree on a new scope of work, and sign a new contract.

Permanent

When your heavy workload appears to be unending, you can advocate to add a new permanent employee to your team. This requires a substantial business case that includes a list of high-priority projects the individual will tackle within the span of 9–12 months. Senior leadership will need to review your case, approve it, and then work with HR to create an opening on your behalf.

Unlike contract workers who can be onboarded in a matter of days or weeks, the process of hiring a permanent individual can take months. Since this individual's livelihood will depend on the company and your ability to provide work for them, this decision must be deeply considered. Do not advocate for hiring someone permanently without a clear, long-term plan for how they will fit into your team. If you're unsure, stick to a temporary or internal solution.

Set the Bar

Staff designers must set the bar for elegant design at their company in direct and indirect ways. They directly model high-quality craft by showing their more junior teammates what fantastic design—and career growth—looks like. Indirectly, staff designers facilitate the creation of high-craft work through a variety of techniques as they work across others.

This means that staff designers are partially responsible for the creation of an environment that ensures others deliver high-quality output. Some folks assume they must micromanage their peers. They swoop into conversations, uninvited, and criticize the work of others without context or consent. They call designs "weird" or say the designs are "not working" without clarification. This generates an environment of fear and displays a lack of trust. You can be better!

Rather than being a micromanager, you can be a facilitator who builds trust and a shared understanding with the designers you lead. Facilitators empower others to be autonomous and only step in to correct course when necessary. They are humble and believe the designers they lead will also have lessons to teach them. They ask questions before proposing solutions, which gives others the opportunity to both explain what they already tried as well as why they landed where they did. This approach creates empathy between you, allows those you lead to release pressure through conversation, and shows you're all on the same side.

CATT'S CORNER
Being a Good Mentor

At some points in my career, I worked in situations where I was directed by other designers. Some of these designers were placed into leadership positions with little training, and their natural inclinations were to assume the incompetence of other designers. These situations showed me the importance of intentionality and building trust with the people you direct.

When I join a new team with the expectation of leading other designers, I put in an effort to learn as much as I can about the people I will lead. These designers will be heavy contributors to the outcomes we're working toward. Proactively opening lines of communication through outreach ensures that we all share the same goals and build trust in each other.

Our introductory conversations also give me the chance to share more details about my role as an IC design leader. In these discussions, I ask to hear their career history—not just their current employer, but also the reason they got into design. This clarifies their motivations and highlights the full extent of their current capabilities. At the end of the discussion, I offer mentorship to each person and schedule additional time with those who want ongoing career discussions.

Mentorship is all about identifying the goals of your mentees and creating a safe space to discuss ways of working toward them. I shift between mentor and coach, sharing firsthand experiences from my career history and then asking questions to provoke improvement opportunities. I usually meet with mentees for 30–60 minutes every 1–2 weeks. As I'm not their manager, we can talk about spicier topics, and I always make it clear that nothing they say will be shared. If I notice a symptom that needs to be communicated upward, I ask for consent and anonymize the feedback.

Some designers decline to be mentored. In these cases, I still schedule regular time to walk through design work. While I'm not directly responsible for designers' career growth, I am on the hook for design quality.

I've consistently found that this approach has made me a more effective leader and deepened designers' trust of senior leadership overall. Several designers I led later told me that working with me was the first time a staff designer made them feel welcomed. This should be the norm, not the exception. ∎

Techniques for Leadership

Use the following techniques to be a more collaborative and inclusive leader.

Share Good Craft

In addition to working in public, staff designers are expected to share work that inspires others to improve on a regular basis. You should present work that spans the four staff designer archetypes from Chapter 1, "What the Heck Is a Staff Designer?," to illustrate the pinnacle of polished design for aspiring junior, mid, and senior designers. Craft quality varies by company, but the more you share work and communicate your thought process, the more others can emulate you.

Some designers create channels within internal communication platforms where others can subscribe to their work. Others regularly bring their work to design critiques to show their peers that no one is above constructive criticism. Experiment to find the methods that work best for you.

Use Situational Leadership

A common quality of micromanagers is a sense of inflexibility. This strangles creativity. Instead, recognize that each designer needs a different approach. Adjust your communication and leadership approach based on each designer's level of experience. You can employ situational leadership, a framework created by Drs. Paul Hersey and Ken Blanchard, to ensure that you spread your energy efficiently across designers.

Situational leadership encourages leaders to give more complex projects and autonomy to those who are confident, willing, and able to handle such responsibility. Junior designers who are either less able or less confident require more support. If they are excited to learn, you can help by directing their energy. And if they are creatively blocked, you must push them toward solutions that push them through their personal ceiling. Learn their motivations so you can propel them forward in challenging times. Tap into the areas of growth they share with you to encourage them even in times of mental blocks or ambiguity.

Share Your Thought Process

Many micromanagers gate-keep the creation of high-quality design by refusing to share the frameworks they learned. This makes it seem as though you're either born with decent taste or you're forever tasteless. Show that taste is earned through practice by explaining how you think when you give design feedback. This will also help other designers understand what matters to you.

When you explain the thought process behind your feedback, you shift yourself from a judge to a mentor. Your communication also improves. So, the next time a designer uses uneven spacing, sit down for a few minutes and illustrate the distribution of spacing so they can see the mismatch for themselves. And the next time a designer uses too many type sizes, count the number they use and ask for a reduction to a specific amount.

Specific, actionable feedback helps designers improve at a rapid pace. Give them the gift of this feedback so they can learn the same frameworks someone once taught you. They'll think like you in no time.

Co-Create Regularly

Micromanagers often "swoop and poop" on others' work—only appearing to comment on designs when it is convenient for them (which is often late in the process) and then hiding away in their ivory tower until the next time they unleash negative feedback. Be different by being present. Participate in the design process early and often. Conduct regular pair design sessions by meeting one-on-one with those you lead to combine efforts and create interfaces in the same file together.

Michelle Kwon, Staff Product Designer at Flowcode, spends 45 minutes every two weeks working individually with each designer on her team. This empowers the designers to work through blocking design challenges with direct support and gives them the direct mentorship most designers yearn for. It also gives Michelle visibility into the progress of design output she is ultimately accountable for.

Spend an average of 30–60 minutes with each designer on your team every 1–2 weeks for maximum effect. Pick an amount of time that works for both parties, schedule a recurring meeting, and give the time back whenever you don't have an agenda. This ensures that you both are connecting regularly.

Pair design sessions are crucial because most designers don't have the option to participate in traditional apprenticeships anymore. They are hungry for mentorship, and they should be—designers who receive hands-on mentorship grow exponentially! The more you help them grow, the more work they can take on with less direction—meaning you can accomplish a lot more together.

Be Direct with Your Direction

Some micromanagers are unclear with their feedback. They send designers on wild goose chases to achieve an unknown ideal. Rather than playing guessing games with your team, share your opinions in a direct way.

When you keep your opinions secret, it frustrates and confuses your collaborators. If you obviously have a direction you want the team to move toward, share it with the team. Make it clear whether you are sharing a suggestion or a requirement.

Guessing is only fun when you're playing games. Build your team's confidence in your leadership by being open and honest about your perspective. Then give them the ability to choose how to move forward. The team will thank you for the clarity as you all move more efficiently.

CREATE DESIGN GUIDELINES

An easy way to make your direction clear is to proactively communicate design principles and recommendations in the form of design guidelines (Figure 7.3). Design guidelines help others understand your thought process by outlining expectations. Typography, layout, color, and other visual design direction should be included. This allows the team to operate independently and ensures that you can use any live discussion time on more pertinent matters.

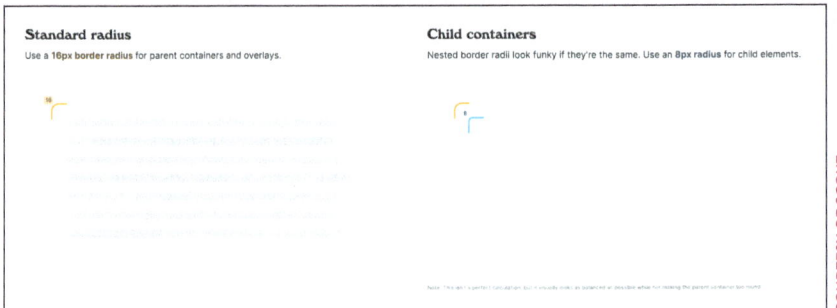

FIGURE 7.3
Design guidelines can reduce the number of feedback loops required by communicating your expectations up-front.

Sometimes, direction isn't enough; teams also need to see examples of real situations to feel confident that they can deliver work at the level expected of them. You can include examples of design work that meets the quality bar (Figure 7.4) in your guidelines. This underscores the principles behind the elegant solutions you expect the team to deliver.

FIGURE 7.4
Example projects can help other designers understand how to make decisions like you do.

Build Each Other Up

The worst micromanagers seek out any loose ends or gaps and use them as an opportunity to tear down their fellow designers' work. Perfectionism can often become toxic, whether it is intentional or not. Give your team positive reinforcement as often as possible and be specific with your feedback.

Positive reinforcement nudges designers toward the direction you want rather than away from the things you dislike. When someone does something you like, let them know! They will feel encouraged to keep doing the things you appreciate, and you will bond more as a team.

Handle Challenges

Delegation is hard work, and even the most skilled individuals will experience the occasional headache. The two main challenges leaders encounter when delegating work are pushback from assignees and time crunches due to their own workload. Navigating these challenges with intention can help you get the best out of your delegation experience.

Pushback

Less senior designers sometimes become intimidated when a staff designer joins their team as a project lead. Some designers feel protective of their autonomy and do not like the idea of being overseen. Freedom can be intoxicating, and they may not give it up without a fight.

This issue can be compounded, depending on the prior dynamics of the team. When someone shifts into a leadership role, it transforms peers into subordinates. Whether or not you intend for your connections to change, they will, without a doubt. Occasionally, jealousy and hurt feelings may surface.

In these situations, you'll need to connect with your fellow designers on an emotional level, so they see you as human and understand your intentions. Build a trusting, collaborative relationship using techniques from Chapter 4, "Nurture Your Relationships." Let them know you're on their side and their walls will come down.

Time Crunches

Busy people think they don't need to be present after they delegate work. After all, you handed over the work so you could focus on other things! But you must find ways to direct a project, even after you hand it to someone else.

Your workload may not always permit you to support other designers as often as you want. Nonetheless, set clear expectations about how much time you do have to be present. Schedule regular 1:1 sessions and be available at design critiques. If you're tired of getting randomly pinged with questions on your company's messaging platform, you can also schedule weekly office hours.

Empower individuals to make decisions without your presence. Use artifacts such as design principles, guidelines, and component libraries to help your team create quality designs in your absence. Only create artifacts that will get used; you can wait until you notice recurring issues or hear the same questions repeatedly.

Debrief

Delegation can help you scale your impact by shifting your workload to involve other contributors. When you delegate your work effectively, you focus your energy on the highest-impact initiatives while creating growth opportunities for others. With the right amount of support on your part, the team's overall work quality will increase at an elevated pace.

There are trade-offs to each delegation method, so set appropriate expectations based on the type of delegation you do. A full-time employee will generally be a better long-term investment than a contractor but may take longer to ramp up. A contractor may be easier and faster to hire than a new permanent employee, but they will not retain the knowledge you share with them.

Delegation invites more variables to your work. Intentionality is the key to successful delegation. Learn what matters to the individuals you assign work so you can appropriately direct their energy and build trust. Create artifacts and frameworks that support your thinking so you can unblock the team without needing to be constantly available.

The more you work across others, the more valuable you will become as a staff designer. This role requires you to take on large, ambiguous challenges that most other employees would balk at. Build your delegation muscle to make a larger impact on the company, thereby justifying your (probably) high salary.

Activity

List 3–5 small or medium projects you could delegate and then explain how you would delegate the work.

Project	Size (S, M, L)	Impact (S, M, L)	Delegation Method	Artifacts to Make

Reflect on the above.

How might delegation of the above projects impact your capacity?

Who would you need to talk to about delegating the above tasks?

What rituals does your team already have in place that would help make the delegation a success?

What rituals would you need to stop or start to make delegation of those projects successful?

Got some strong candidates for delegation? Grab 30 minutes with your manager and discuss the idea using the influence formula (observation, proposal, outcome).

CHAPTER 8

Show Your Value

Why Humility Fails	170
Manage Your Presence	172
Manage Up	177
Debrief	188
Activity	188

Staff designers work across many people and are often responsible for high-visibility projects. This can lead individuals in this position to think their work will speak for itself. In fact, I used to be one of those individuals. Early in my career, I put my head down and worked my butt off. But because I didn't talk about the work, people above me didn't know what I was doing.

Companies have many moving pieces, and senior leaders are busy. They usually don't have the bandwidth to keep up with what people are doing unless those individuals proactively communicate. A staff designer's position is critical—and usually expensive. Therefore, it's even more important that a designer at such a high level continuously justifies their value by regularly underscoring their impact.

In addition to communicating value, designers must proactively create awareness about their work in order to be perceived as leaders. They must permeate the minds of their direct collaborators and those above them, building confidence in their abilities and showing their usefulness. When successful, the designer will be consulted on a continuous basis and seen as a subject matter expert.

There are two ways in which a staff designer can communicate their value and manage how they are perceived. The first method is intrinsic: a designer can focus internally on how they carry themselves and work to come across as more of a leader. The second method is extrinsic and centers around managing the expectations of other senior leaders (as shown in Figure 8.1). By taking an intrinsic and extrinsic approach, designers can align the way they perceive themselves with the way they are perceived by others.

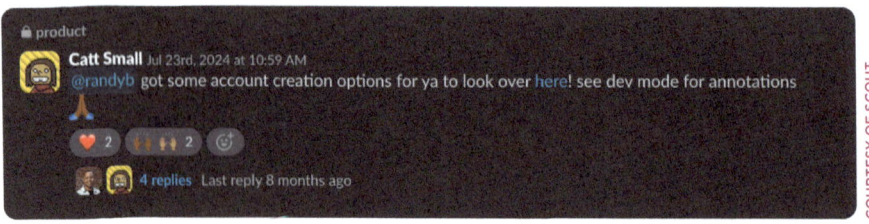

FIGURE 8.1
Visibility doesn't have to take lots of work—a short and simple message in a team channel can provide lots of value.

IN THE REAL WORLD

FIONA ROLANDER

Fiona Rolander is a Staff Product Designer at Dropbox. She has worked at Dropbox for over six years. She joined at the senior level and then shifted to management after two years. After three years, she shifted from a senior manager role back to hands-on design and was assigned the staff level.

Fiona's management experience helped her understand what information her own manager would benefit from most. "When you get to a certain level, your manager is the GM. They don't have time to focus on all the little things." She believes staff designers should give high-level visibility into the status of their work, set expectations about timing, and flag concerns. "The focus has been more about the impact my work is having."

In comparison to the senior level, there is a major difference in how Fiona communicates with her manager as a staff designer. "As a manager with more junior reports, I needed to know what they were doing so I could help them triage and prioritize." As a staff designer, the expectation is that she can self-direct the majority of the time. Over time, she has shown her manager that she is a capable individual who doesn't need tactical support. "As long as you are generally operating as expected, then there's no need for micromanagement."

Being a present member of the team is a critical part of Fiona's work as a staff designer. "Speak up. Unmute during that Zoom call with 200 people. Post in that big channel. You're never going to influence if you're invisible. And in a remote world, girls have to use the tools available." Rather than shying away and hiding, Fiona takes up space and ensures that she is perceived as a leader by the team.

Why Humility Fails

For many individuals, the idea of talking about accomplishments is uncomfortable. In many cultures, humility—or the act of not discussing your favorable qualities—is seen as a signal of virtue. This was how I was raised, and like many, I believed that the quality of my work would speak for itself.

While humility and self-awareness are great traits to have, the workplace requires a different approach. It's important to convey confidence in your skills in the modern professional world. If the person doing the work isn't confident in their abilities, their employer won't be able to trust them.

The people who speak proactively about their work also get rewarded because they create awareness about their efforts. Because they are top-of-mind, they get the best assignments and have the best outcomes. This makes self-promotion a crucial skill for getting access to the projects that shape a staff designer.

Meanwhile, the people who share nothing about their work are forgotten, and their results appear to be nonexistent. Kritika Kushwaha, Staff Product Designer at Asana, used to struggle to broadcast updates. "I was really focused on getting results and figured the work would speak for itself," reflecting the common assumption that a manager will know what a designer is up to without any effort on the designer's part.

Over time, Kritika learned that her manager wouldn't be aware of the extent of her contributions unless she shared more detail. "I started to let my manager know up-front." Kritika began regularly writing down bullet points like the ones in Figure 8.2 to summarize her achievements. "It's one thing to do great work, but if you want to grow in your career you also need to make sure your impact is seen by the right people, and that means being intentional about how you share it." This simple change helped her build the documentation necessary for a promotion.

Awareness helps designers get promoted, but it can also be a tool for education. When a designer shares how they work, they give others insight into their own design process. This amplifies the flow of information between individuals and can provide visibility that leads to more efficient work.

- Jumped on to help with GDPR privacy policy work with little warning in advance
- Made contributions that shape the company's culture with efforts such as creating an experience that lets users block unwanted messages
- I have kept the analytics redesign project going despite not having a consistent engineering team and full-time PM for six months
    - When we didn't have a full team I helped to define my own work, prioritize tasks for myself and the team, and break down a very complex project into a set of user-centered ideas that could be validated through testing
    - To ensure that the team started on the right foot, I organized a team norming session to make sure expectations across roles are clear
    - I keep team meetings productive by requesting an agenda and a goal, plus help to make sure meeting facilitators are set up for success
    - I help keep us moving forward and look ahead to remove impending roadblocks
    - Put work in front of customers in 1 month or less, at least half the normal time
    - Releasing features to customers in September, less than 1 year – which is way lower than before
        - While the project initially started in November, we didn't get a full dev team until May – meaning we're launching in 4 months

FIGURE 8.2
A straightforward list of bullet points can help your manager keep up with all your accomplishments.

When sharing is viewed as a form of education, humility loses its appeal. Designers should consider speaking about their work through the lens of helping others. Everyone should be a team player and toot their horns!

> **TIP HANDLE IMPOSTER SYNDROME**
>
> *Impostor syndrome*, or the feeling of *not* being qualified for a role despite getting hired, often makes self-promotion challenging for designers. If you struggle with impostor syndrome, I encourage you to consider what makes you feel uncomfortable about sharing your work. What would be necessary to help you discuss the impact of your work authentically? How might you make it feel less like bragging and more like sharing valuable information with your team?
>
> Communicating the status of your work can be a factual exercise. If you're putting in the work, people should know. You don't have to be a braggart to get ahead, but you do have to share proactively what you do with others.

Manage Your Presence

The first lever of perception management is presence. How a designer shows up matters as much as how often they communicate. To be true leaders, designers must invest time and energy into bringing the best version of themselves to every conversation.

There are several components of presence:

- **Demeanor:** The way a person appears
- **Succinctness:** How indirectly or directly a person communicates
- **Candor:** The level of honesty with which a person conveys their thoughts

The word *presence* might sound familiar. *Executive presence* is a concept that determines whether or not people are confident in the leadership skills of an individual. According to HBR (*Harvard Business Review*), the traits that comprise executive presence change over time (see Figure 8.3). For example, forcefulness was desirable in 2012 but is no longer seen as a valuable trait for leaders to have today.

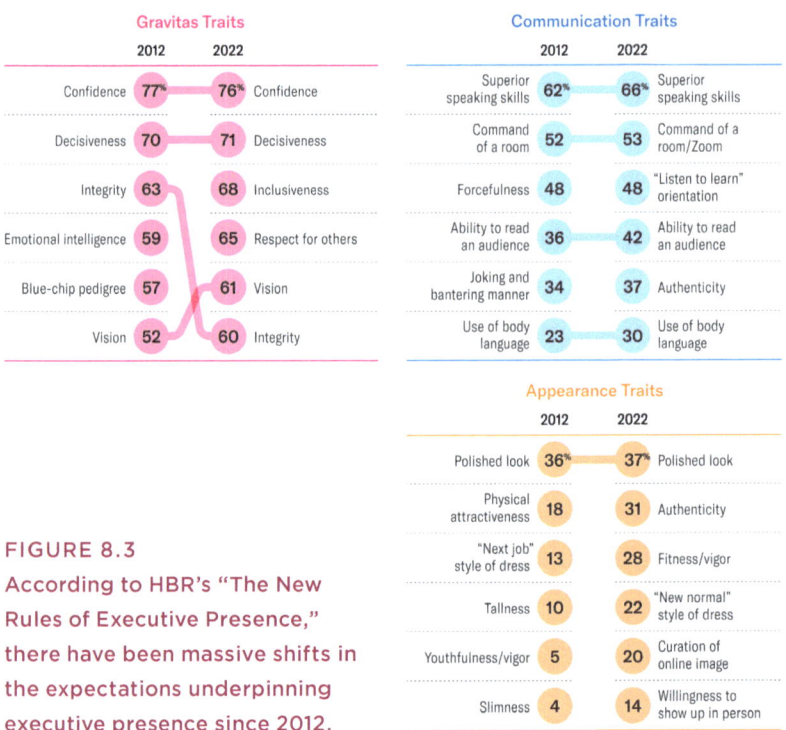

FIGURE 8.3
According to HBR's "The New Rules of Executive Presence," there have been massive shifts in the expectations underpinning executive presence since 2012.

However, some traits have staying power. In both 2012 and 2022, leaders were expected to have confidence, decisiveness, integrity, and vision. They are also still expected to command a room, read the audience, and present with superior speaking skills.

It is crucial to build executive presence, as staff designers are also senior leaders. They may not have direct reports, but they are expected to make a huge impact. Each component of presence builds a picture of an individual and determines how they will be perceived.

Many people operate subconsciously, missing all the ways in which they subtract from their own presence. The goal is to be intentional in all regards. A designer who manages their demeanor well, communicates succinctly in the active voice, and understands when to be more or less candid will elevate their presence to that of a senior leader.

Demeanor

A staff designer can choose to be perceived as more or less authoritative by adjusting their demeanor. If a designer is in a situation where they need to engage with other leaders, they might *increase* their authoritativeness to be seen as a peer. Around other hands-on designers, they might *decrease* their authoritativeness to make the atmosphere feel safer.

Demeanor impacts trust. By adjusting the way they appear, a designer can either be seen as an equal, above, or below the people they interact with. Some ways a designer might impact demeanor include body language, such as facial expressions and posture. A classic confident posture is the akimbo pose (shown in Figure 8.4). Individuals with relaxed or self-assured facial expressions are more likely to be perceived as welcoming and confident.

At some companies, clothing can also have a major effect on one's executive presence. Individuals at small companies can often dress down without negative repercussions. In larger, more corporate environments, a shift from T-shirts and sneakers to blouses and formal shoes can signal that an individual commands respect.

FIGURE 8.4
Experimenting with the way you bring yourself to work, even just a shift in your posture, can result in a boost to your confidence and impact your presence.

The value of demeanor adjustments depends on the context of the company itself, as well as the relationships between the individuals involved. At a small start-up, a designer will likely have to make less of an effort to be seen as competent because there is less presentation work to do—the senior leadership team is closer to the team and usually has more awareness of their day-to-day efforts. Conversations with senior leaders at a large company might be rarer, so each interaction has a major effect on their impression of hands-on designers.

> **TIP DON'T BE LIKE JEKYLL AND HYDE**
>
> The core personality of an individual doesn't need to shift just because they adjust their demeanor. A simple change in body language can go a long way. Adjust your demeanor based on your personal and professional goals.
>
> While some individuals can get away with not having to adjust their demeanor, I believe all designers should deeply consider how they want to be perceived by their peers and leadership. It's important to intentionally convey messages through appearance and nonverbal cues. These seemingly small attributes make a big difference.

Succinctness

Words can help a designer own their agency. Language also impacts how individuals are perceived by others. Sarrah Figueroa, Staff Product Designer for a major content publishing platform, recommends designers "pay attention to the types of words" they use in different situations. "Who's in the room when you're using them?" When surrounded by peers, staff designers have the leeway to speak any way they want.

When executives are in the room, Sarrah suggests designers "be a little bit mindful" because language has a heavy impact on their perception. One major way to adjust language when speaking in front of senior leaders is to shift from passive to active voice (as shown in Figure 8.5). Passive voice is indirect and verbose; active voice is direct and succinct. A designer can build their audience's confidence by using action-oriented language.

~~was designed~~ → I designed

~~was validated~~ → The team and I tested

~~is not done~~ → I will do

FIGURE 8.5
Small shifts in language have a large impact on clarity.

Instead of saying an experience "was designed," a designer can say they "designed the experience" to perform in a certain way. If a design "was validated," the designer can say they "tested it" and explain the result. When other teammates are involved in the process, the designer can credit them by name or refer to the broader team itself. This can also apply to situations where the team has yet to perform a certain task—instead of saying the task "is not done," the designer can say they "will do" the task and set expectations around timing.

A shift to active voice feels like a subtle change, but it can have major effects on the perception of a designer's skillset. By taking ownership of work and giving credit to others, a designer using active language takes the guesswork out of communication. As senior leaders are busy people, this clarity is often much appreciated.

Designers must learn to recognize when they are about to speak in a situation that would benefit from the use of active voice. This change takes time to feel natural. Practice goes a long way, so designers who want to improve at using active language should try speaking actively about a project alone. Repeat the process until it becomes more organic.

> **TIP MISTAKES ARE NORMAL**
>
> Communication is a skill. It's okay to make mistakes along your journey to become a more active speaker. If you find yourself falling back into passive voice, try to correct yourself and switch to active voice in the moment. I have done this lots of times. It's important to avoid drawing too much attention to the mistake itself; instead, just swap words quickly and move on. People will only notice if you put a spotlight on it!
>
> This is also true for other situations where you're presenting information. Mistakes happen; perfect is the enemy of good. The audience usually wants you to succeed. So, if you make a mistake, give yourself grace. If necessary, clarify what you meant using as little energy as possible—plus as much confidence as possible—and keep going.

Candor

Candor, or the ability to express oneself sincerely, is another factor that can impact a designer's executive presence. Directness feels more authentic and approachable, but it can also trigger feelings of discomfort or frustration in leaders. Abstraction can appeal to other senior leaders but feel opaque to direct collaborators along with junior members of the team.

A staff designer modulates their level of candor based on the audience to ensure they are communicating more effectively. By reading the room, a designer can identify the circumstances in which more candor would engender trust—and, conversely, when less candor would be more effective. This is an important part of the designer's political toolkit.

Many designers struggle to modulate their level of candor, instead opting for a more heavy-handed approach. This can be harmful to a designer's career, as other senior leaders will be concerned by what might seem like an inability to handle delicate situations. A large part of leadership is knowing when to hold your tongue and when to spit fire.

> **TIP DETERMINE WHO'S TRUSTWORTHY**
>
> Some senior leaders appreciate candor more than others. Don't come out the gate expressing strong opinions, as this might offend some people. Instead, start communicating with restraint until you can gauge the environment.
>
> Build a mutually beneficial relationship that centers curiosity by asking your leaders questions and getting to know what they care about. Their answers will help you understand who values directness and who prefers less candor. Over time, you can show your hand and speak more directly with those who appreciate sincerity.

Manage Up

Designers' relationships with their managers are top-down for most of their early career. This relationship continues as the designer matures and becomes senior. They continue to function similarly until a shift occurs and the designer reaches the staff level. They are now expected to self-direct and manage their manager's expectations.

This change catches many designers off-guard. I've had lots of conversations with individuals in my staff designer course about reframing the relationships with their managers. Designers are not set up to make this transition without encountering a handful of hiccups. No one directly communicates that the change is coming until it's too late, and then they punish the individuals who don't keep up.

There are three components involved in managing up:

1. **Set expectations.** Define communication agreements before each project.
2. **Share regular updates.** Provide consistent communications as a project progresses.
3. **Broadcast impact.** Celebrate the wins and takeaways of a completed project.

Designers must invest equal amounts of energy into each component to ensure they are working efficiently and creating continuous awareness of their value. Setting expectations without following through results in a fall-off. Sharing progress without expectations can result in misalignment. Sharing impact without earlier communication can surprise or frustrate leaders. These three actions produce a continuous stream of information that shows an individual is deserving of the staff title by ensuring a team's success.

Set Expectations

At the beginning of every project, staff designers set clear expectations with their stakeholders up-front using artifacts like the one shown in Figure 8.6. Expectations include agreements regarding who must be involved. The team must also decide on the formats in which updates will be shared with stakeholders, and how often the updates should be communicated.

Team

Name	Role	Responsible for
Alexis Brown	Product Manager	Project roadmap, prioritization
Sarah Jones	Product Designer	Driving the design
Maria Garcia	Senior Engineer	Technical feasibility, build
Ben Miller	User Researcher	Gathering user feedback
Yoonju Kim	Director of Product Design	Providing feedback
Kevin Ntim	Director of Product	Providing feedback
Emily Lee	VP of Engineering	Providing feedback

Meeting Rhythms

Name	Purpose	Frequency	Attendees
Standup	Updates on project status	Weekly on Mon	Alexis, Sarah, Maria
Design Crit	Review in-progress design work and capture feedback	Weekly on Tues	Sarah, Ben, other designers
Team working session	Working session with PM and Eng	Weekly on Weds	Alexis, Sarah, Maria, Ben
Leads update - Async	Share project status updates	Weekly on Fri	Alexis, Sarah, Maria, Ben, Yoonju, Kevin, Emily
Leads review	Gather feedback and receive direction	Once every 2 weeks on Fri	Alexis, Sarah, Maria, Ben, Yoonju, Kevin, Emily

FIGURE 8.6
Defining the team's roles and aligning on the communication frequency helps everyone operate efficiently.

At this stage, the team can also agree on expected outcomes of the work, including success metrics. Like other situations, the same hierarchy of metrics applies during this alignment conversation. The more a team can tie work and communications to quantitative, positive measurements, the more leadership will value their efforts. Quantitative, negative metrics will be less compelling, and qualitative metrics will only have sway with the most customer-minded approvers.

Aligning on these preferences up-front gives the team time to identify and address any misalignments with approvers before the work begins. It also serves as a contract that can be referenced later. This approach ensures that everyone communicates their needs in advance and has time to express their communication preferences.

> **TIP DOCUMENT YOUR AGREEMENTS**
>
> After negotiations with the approvers, the team can ratify their agreements by documenting them in a project brief. This will ensure that everyone remembers the original agreement. If a communication method or frequency needs to be changed after the project has kicked off, the team must update the documentation to reflect any adjustments.

Communication Methods

There are many ways to communicate the state of work to approvers and observers. Some methods require more active effort from those driving the work, while others require more energy from those who seek the information. Designers should aim to work with the rest of the driving team, approvers, and observers to maximize impact while minimizing the team's effort.

The key categories of communication methods are *synchronous* and *asynchronous*. Asynchronous methods usually require less effort from the driver but take more energy for approvers and observers to consume. Synchronous methods often require more effort from the driver, who must present the information in different live conversations, but often helps the recipient understand the information with less effort on their part.

Asynchronous communication methods include:
- Posts in a team channel on the company's preferred messaging platform
- Emails to specific teams or sets of individuals
- Recorded videos
- Updates to a section within a particular document
- Status updates posted to a specific project in the team's preferred project management tool

Synchronous methods include:
- 1:1 meetings with relevant individuals
- Small group meetings with 3–9 people from relevant teams
- Large group presentations to audiences of 10+ people

The team can align on a mixture of asynchronous and synchronous methods as necessary to ensure that the right information is delivered to the right people. For example, the team might decide that the primary communication methods are asynchronous, but they will host a small synchronous meeting at a manageable cadence and present to a large group once at the end of the project. This maximizes value by concentrating the majority of the team's energy on actual project work and saving the heavier communication work for the end of the project.

Communication Frequencies

In addition to communication methods, teams must align with approvers regarding how often they will provide updates. The cadence at which a team shares status updates and requests input will impact the amount of energy they are able to expend on the work itself. A staff designer works with the rest of the driving team to align on a proposed cadence that maximizes their time to focus on the work at hand while also giving approvers the awareness they need to feel confident that the work will be completed on time.

Every approver has an opinion regarding how often they need information. Some approvers need to receive information more often so they can report on the area they are responsible for. This is especially common at large companies. Jess Dale, a Senior Manager with extensive experience managing staff designers, lets reports know: "I'm going to be in meetings with leadership within my product group—or other product groups" and accurately representing their reports' work is critical for being able to properly underscore "the ways that design impacts the business or the product overall."

The more a team proactively communicates the information an approver needs, the more rapport they will build. Designers and driving teammates must be aware of any forums where the status of their work will be referenced. By understanding where the information will be used, the team can ensure they communicate at a cadence that supports the updates of those resources.

Share Regular Updates

As a project progresses, the team will need to provide consistent updates to their approvers. The designer leading the project will be expected to communicate major design decisions, sometimes collaborating with their PM and engineering partners to walk through the constraints that influenced their decisions. A staff designer tells a clear story by communicating decisions at the right altitude and facilitating the conversation, so the team receives the necessary guidance from leadership to proceed.

Senior leaders are high-level thinkers who switch contexts at all times of day. They have dozens of meetings every week, and—depending on the size of the company—have to pay attention to multiple active projects at the same time. Staff designers must communicate accordingly to ensure their approvers have the context required to move projects along at a reasonable pace.

There are three types of information to include when speaking to an approver:

- **Decision:** What the team will do moving forward
- **Context:** The inputs that led to the decision
- **Resources:** Further reading and other information, such as an appendix

An example ratio of these three types of information might be to focus 75% of the discussion on the decision, 20% on the context, and 5% on resources. This formula ensures that the conversation centers on the decision that approvers need to greenlight. As shown in Figure 8.7, it can be customized or rearranged based on the needs of a particular leader. This is important to keep in mind because some leaders prefer the decision to come after some context, and others want to see more context to feel confident in the decision.

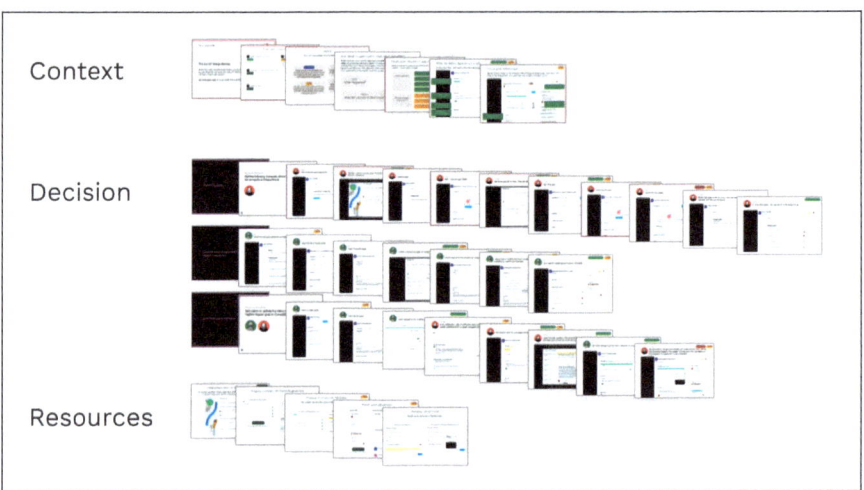

FIGURE 8.7
A designer can customize and rearrange the formula to best suit their organization's communication needs.

Decision

It's important to announce the top-level design decisions to approvers and observers because it helps the team get eyes on the topics that matter most. Whenever possible, the coverage should be focused on the decisions that the project's approvers either need to agree with or unblock. The team can also broadcast decisions that were made in conversations that preceded the update.

Teams can avoid getting bogged down in the nitty-gritty details by focusing decision-making conversations on the big picture. It's not efficient to converse about small-scale micro-interactions, nor do most leaders care to see every single detail of every use case. Most leaders want to be confident that the majority of customers will have a positive experience and predefined success metrics will be met. Staff designers concentrate on delivering this clarity.

> **TIP INCLUDE NEXT STEPS AND OPEN QUESTIONS**
>
> In addition to the decision itself, a designer can provide next steps to give leaders confidence that the team will continue moving forward. This also gives leaders a chance to redirect the team if they disagree with the planned actions. Highlighting big questions that are still open can also give leadership insight into the kinds of support the team needs to unblock next steps. Whenever possible, communicate this kind of information so leaders know how they can keep the team moving quickly.

Context

A staff designer predicts the questions that leadership will ask and preemptively answers them. They must walk through the thought processes and limitations that resulted in the decision being shared to help approvers understand why the team landed where they did. This information ensures that any and all detractors will agree with the team's decision. The context presented should include key forks in the road, along with a project timeline and any other high-level information that might convince someone to consent to the decision.

The context should be very focused so that any potential questions about the key decisions are answered. For example, a compressed project timeline might explain a trade-off the team had to make. Or certain engineering constraints might resolve concerns about a critical user flow.

A staff designer knows the value of being concise. If the conversation is not focused and intentional, leaders will start going down rabbit holes and land in places that potentially counter the team's decision. All context should be relevant to the conversation, and nothing extraneous or distracting should take up space.

Resources

Some approvers want to see the thinking behind the decision that was made (hence, they are especially scrutinizing), so designers should provide links to further reading. It can also route all who may be concerned to any relevant files and documentation they might need to reference. This should take up the least amount of space possible to avoid distracting everyone from the major decision.

Broadcast Impact

Once a project is completed, the team must monitor results and then share their impact. A staff designer is a leader, so they will need to help celebrate the work of their team. This might be uncomfortable for some individuals who are not used to talking about their accomplishments. But sharing the impact of hard work is crucial for distributing knowledge and building team morale.

An excellent project completion announcement includes three components:

- **Outcome:** The metric or qualitative impact
- **Output:** The customer experience that enabled the change
- **Process:** How the team created the experience

These three components ensure that other teams can learn from a change made to the customer experience. These components also provide valuable historical documentation in case future teams need to reference the work. With a little bit of effort, a designer and their collaborators can assist others in replicating their win.

CATT'S CORNER

Share Successes and Failures

You might be tempted to only share success stories. Failure is sometimes viewed as an embarrassment. In reality, failure is a part of life. We all fail all the time.

It's critical to view failure as a necessary part of growth. People make decisions based on what they know at the time. It's impossible to predict the future. Despite this reality, people often feel ashamed of failures and blame themselves for making mistakes—even if those so-called "mistakes" were informed decisions based on valid hypotheses.

Product development failures are common. We shouldn't avoid or ignore the results of our own actions. By reflecting on and sharing the decisions that led to a negative outcome, we can prevent others from making the same mistakes. The lessons you learn can help another team in the near term or the far future.

I always recommend that teams perform a postmortem or "five whys" (or 5 whys) session (Figure 8.8) to uncover the root cause of a failure. Five whys is a technique created by Sakichi Toyoda for use within the Toyota Motor Corporation. During the exercise, teams identify a problem and interrogate the reasons behind it. Each time they identify an answer, they question the answer until they have reached five degrees' worth of answers. While it may not always identify the root cause of an issue, it can help to highlight some of the biggest hurdles the team faced.

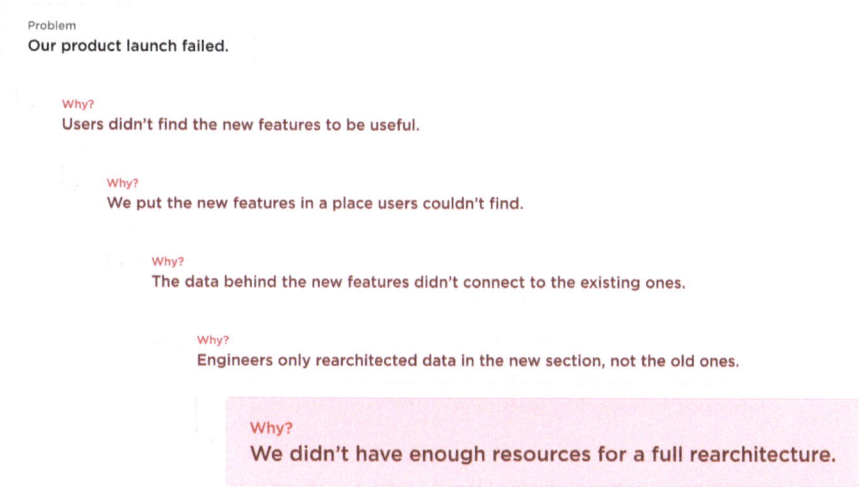

FIGURE 8.8
The five whys is an example of a technique you can use to help identify the root cause of issues that your team faced during a failed project.

Once the team has reflected, we share what we learned. In one situation at one company, the root cause of a failure was the lack of design involvement in data architecture decisions. This gave me and other designers proof that we should be more involved in strategic conversations.

In that same situation, the team also realized we scoped our problem area too small, due to a fear of rearchitecting the whole product. We had wanted to start small because the company hadn't given us many engineering resources, and we didn't have the confidence to ask for more substantial support. Since our experiment failed but the initial problem persisted, we were able to use the failure as data to inform a request for a larger engineering team. Over the following year, we built out a successful solution that properly resolved our customers' complaints.

I believe that failure should be discussed regularly and shared widely. I've even presented my own failures to the public at design conferences around the world! You might not present your failures in public like me or even at an internal all-hands meeting, but perhaps you might find it worthwhile writing up some documentation that other teams can reference. Every time I've shared what I learned, people have been grateful for the insight. Help other people by owning and reflecting on your failures. ■

Outcome

The outcome of the project is the measurable effect the work had on the customer experience at scale that ideally has a direct impact on the business. Staff designers communicate the outcomes of their design work because it clarifies the return on investment (ROI) of the energy they expend.

The ROI, or the business value added for the energy invested into an effort, is key to showing value and prioritizing future efforts. Prarthana Johnson, Head of Design of several core platform experiences at Atlassian, says staff designers can show ROI through qualitative customer feedback or by increasing certain metrics. A successful staff designer is highly collaborative and influential, impacting the business heavily—"and that is seen visibly through changes in the roadmap." The staff designer will help lead the team to execute and is therefore able to claim that they contributed to the impact.

A team can only track the outcome after a project is complete if they agree on the metrics they believe they can move at the beginning of the project. If their hypothesis is correct, Prarthana believes that is the ROI of the team's work—including the design investment. "You should be able to point to those outcomes and say, 'without a doubt I led that, and I did that.'"

Output

The output of the work is the customer experience that was delivered to reach the outcome. The output can be expressed as a link to the live experience, a series of screenshots, a video recording, or a design file with an experience walkthrough. This helps people who were not on the team understand the change to the experience.

A staff designer provides useful documentation now and thinks of ways it might be used in the future. For the sake of posterity, they might include some examples of the old experience and compare it to the new one. This helps the uninitiated identify the difference and has the added benefit of making the designers on the team look good. "Before-and-after" examples are especially important to include when a team has delivered a major redesign, as folks will want to visually compare the old version to the new one.

> **TIP** **PREPARE DOCUMENTATION EARLY**
>
> Working on a redesign or another kind of customer experience improvement? Grab screens and a recording of the old experience before the new one launches. To get ahead of this, do an audit early in the project so you have the screens available for later use. This is one of many examples of how artifacts can be reused in the product development process.

Process

The process outlines the work that was done to arrive at a certain outcome by delivering outputs. By documenting the way the major design decisions were made, a staff designer can model excellent craft for others and help the design team level up their systems thinking. This is an important attribute of being a designer at this level; staff designers don't just deliver great work themselves, but they also help others improve by showing what great work looks like.

In an initial announcement, the process described might be limited to the amount of time the project took and who contributed to it. However, providing a link to a design file with the solutions considered can often help others dig into the work more. Past artifacts can also serve well in this situation. For example, if a designer recorded video updates for each decision and organized them into one place, they could easily link to the repository in their announcement.

Part of sharing the process involves giving credit to all who shaped the project. Great leaders keep track of everyone who directly influenced them so they can applaud them at the end of a project. Sharing the spotlight is a great way to practice useful humility while also tooting one's horn. These shout-outs serve as moments of solidarity and not only build morale but also increase the team's trust in the person giving the credit. It also underscores the authenticity of the individual by showing they are a team player.

> **TIP** **SHOW YOUR FUTURE THINKING**
>
> The work doesn't have to stop just because the project is over. Document and share future ideas for additional improvements so the team has ideas for the next steps. This might also inspire others to take on ideas that your team doesn't have the capacity to deliver.

Debrief

Talking about impact is an important part of the design process. A staff designer knows they must communicate their value constantly and proactively. From the beginning to the end of an initiative, there are opportunities to set expectations, keep teams informed of major decisions, and communicate the ROI of the work. Designers must step up and take hold of each opportunity to invest in their presence as a leader on the team. Humility is wonderful, but a real leader knows the importance of a little celebration!

Activity

Reflect on your presence and the environment at your current or most recent employer. Answer the following questions.

How do leaders at this company present themselves? (Demeanor, Succinctness, Candor)

What are three things you could do to increase your executive presence?

Next, take a project you're planning to work on or are working on at the moment. Document the current communication methods and frequencies of communication for this project.

What is the name of the project?

What are the ways the impact of this project will be measured?

What are 1-3 forums in which your manager might have to share updates about this project?

Communication Method	Expected Outcome	Frequency	Attendees

If you're able to fill the last table, that's great! Got nothing? You need to communicate more proactively. Come up with a few ideas and then discuss them with your team.

CHAPTER 9

Keep Your Career Fresh

Change Companies	193
Go Beyond Staff	197
Switch Roles	199
That's All, Folks!	208
Debrief	211

A 2025 study of over 100,000 workers across various industries showed that career satisfaction often dips for mid-career, highly skilled workers. Many staff designers have been working for over 10 years, meaning their careers have matured substantially. There's a natural ambiguity that comes with reaching this level, and it can leave you feeling unsure what comes next. That ambiguity can be exciting if you want to design your own design career. But it can feel like a curse if you're beginning to feel disengaged.

There are lots of directions you can go to keep your work interesting, as shown in Figure 9.1. You can change employers to continue at the same level of work while refreshing your environment. If you want to increase your altitude and impact, you can keep leveling up beyond the realm of the staff level. Or you can shift into an adjacent role using many of the skills you've acquired.

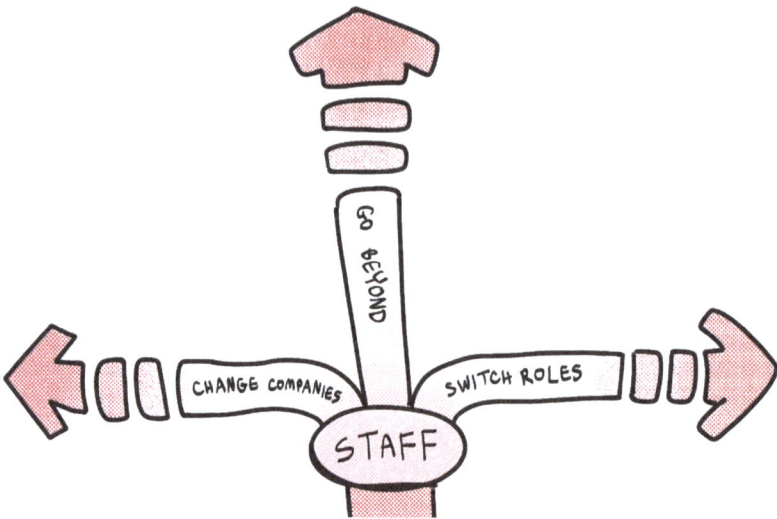

FIGURE 9.1
You can make a lateral move between employers, keep growing, or shift into another position.

Regardless of how a designer decides to continue their growth journey, what matters most is the self-awareness they bring and grace they grant themselves during the process. No one's growth is linear, and every designer's path looks different. New learning opportunities will present themselves to a designer as time passes. Each opportunity will signal what replenishes or subtracts their energy, and these signals will be the guide toward future growth.

Change Companies

There will come a time when you need to look for a new role. Unlike the senior design level, staff designers are more likely to define and execute on large scopes of work. Therefore, the qualities a designer needs to display are different. A designer interviewing for a staff position must show that they have a track record of driving critical projects with a high level of ambiguity and working across multiple teams to create usable experiences that get results. Each of these signals must be proactively communicated so the hiring team can be confident you're a fit for the role.

Drive Critical Projects

A staff designer will likely lead design for some of the highest-impact intiatives at the company they are interviewing for. Therefore, anyone interviewing must show a history of working on mission-critical projects. An example of this is shown in Figure 9.2. Whether you've shipped an important redesign or led a 0-to-1 launch of a new product area, what's important to explain is why your past employer put you on the project. Clearly explain the customer pain and the business problem so the hiring team can understand how you frame challenges.

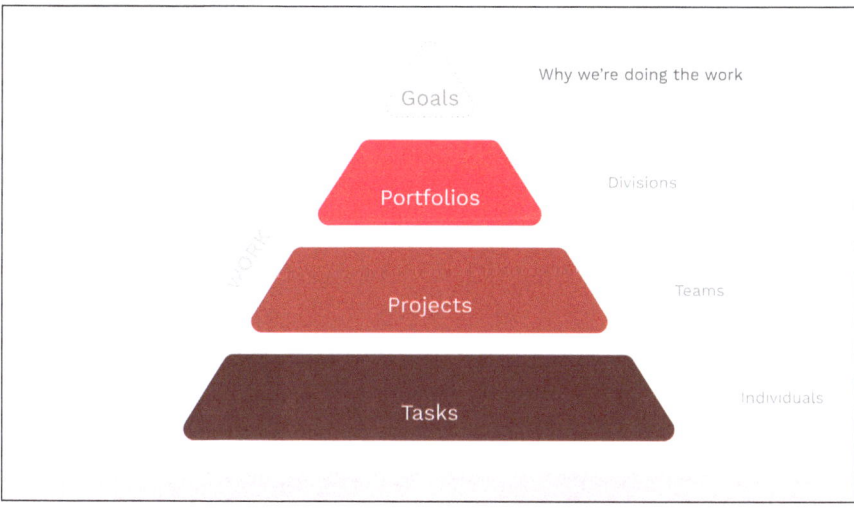

FIGURE 9.2
This slide from my portfolio presentation explains the business-critical problem my former employer aimed to solve with the project I worked on.

Navigate Ambiguity

The more experienced a designer becomes, the more ambiguous and complex the problem spaces they will encounter. Candidates interviewing for a staff-level position must show that they can create momentum in times where the complexity might overwhelm others. As shown in Figure 9.3, designers with a robust toolkit will be able to explain how they continuously used design techniques to keep the team moving toward a solution for a particularly crunchy problem. This will show hiring managers they are able to operate autonomously when necessary.

FIGURE 9.3
In portfolio presentations, I shared ways I unblocked decision-making such as concept testing after a design sprint.

Work Across Others

Staff designers are often responsible for products or areas that intersect with other teams' work. Someone interviewing for a staff design role must show that they are able to define and evangelize a shared perspective across multiple designers and teams. They must also demonstrate an ability to ensure their design direction is implemented as planned. An easy way to do this is by highlighting situations in which you had to collaborate with other teams or individuals to implement a change to a valuable customer experience. Don't be afraid to credit others in your presentation, as shown in Figure 9.4. The story should center on the candidate, of course, but calling out the people who helped along the way can show how they operate at scale.

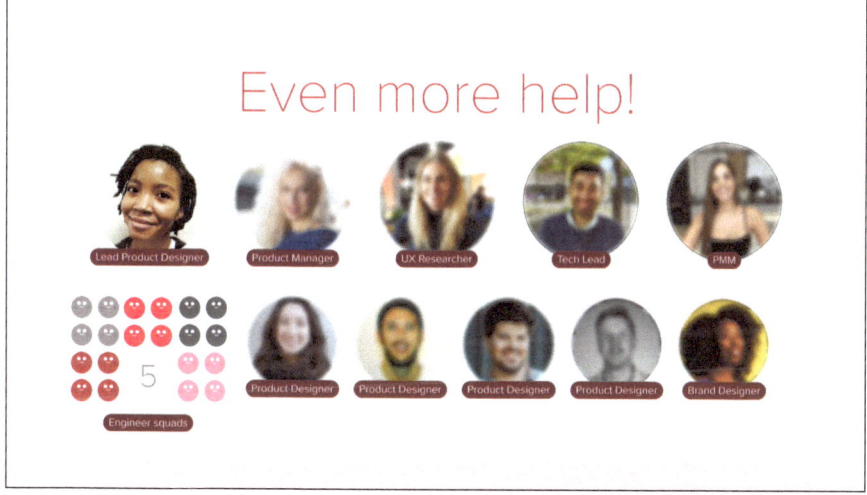

FIGURE 9.4
Help the hiring team understand the scale of your impact by crediting the others you worked with.

Create Usable Experiences

Since staff designers are, in fact, designers, they will still be expected to create high-quality experiences for the people who use their company's products. During the interview process, a candidate must show how they integrated the customer's pains and needs into their work (as shown in Figure 9.5). There should be a clear throughline from customer feedback to design decisions. This shows that you create products and solutions customers actually want. Since a staff role is ultimately an investment, teams want to be sure you'll pay off.

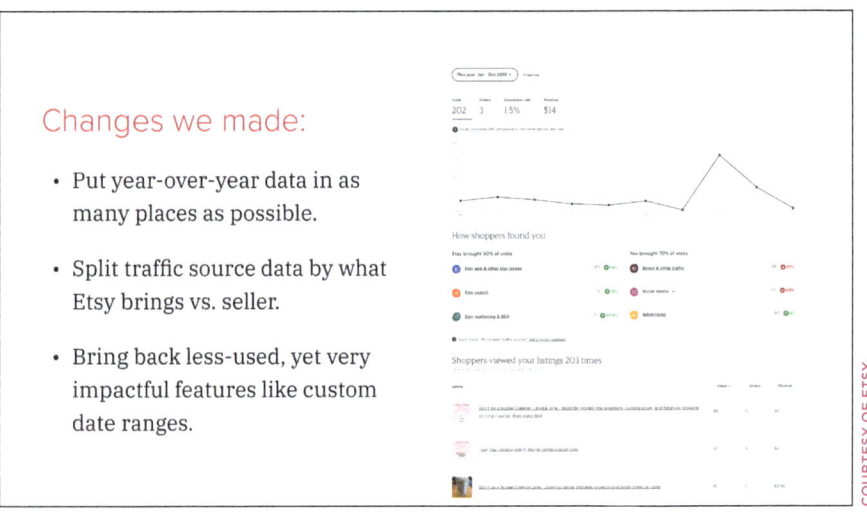

FIGURE 9.5
In hiring conversations, I always call out the design decisions I made based on customer feedback.

Get Results

Many designers focus on qualitative impact rather than return on investment. A strong candidate for a staff designer role presents their business impact during hiring conversations (Figure 9.6). As mentioned in earlier chapters, quantitative metrics will be most influential in this situation as well. If you left an employer without capturing data regarding your impact, you can always list the metrics the team planned to track and clarify the situation.

Launch and Metrics

The new experience launched to 2+ million sellers in October 2019.

GOAL KPIS

- Monthly active Stats users
- Google Shopping MAUs

NEXT STEPS

Continue building toward North Star experience and invest in showing recommendations based on data

FIGURE 9.6
Present the results of design projects to show the hiring team you mean business.

Go Beyond Staff

Reaching the staff design level is a major achievement because it often corresponds to a high level of ownership and autonomy. As a company scales, staff designers will continue to build further experience and eventually progress to an even higher level on the individual contributor path. These levels generally have a title like "senior staff" or "principal." For the sake of clarity, I will refer to the role above staff as "principal."

A principal designer is usually expected to provide direction for new business initiatives and unlock major opportunities that make it possible for their employer to multiply their growth. Tom Takigayama, Principal Experience Designer at Justworks, described his responsibility as "helping to connect experiences across different teams while connecting the business to those experiences through design." Principal designers usually operate at a stratospheric altitude, orchestrating outcomes over 6–18 months, depending on the project and size of the organization.

While every organization has different expectations for a designer at this level, a common thread seems to be that a principal designer finds innovative solutions to the most business-critical customer problems. These individuals often float between areas and provide impact wherever necessary. Due to this floating nature and the scope of work, the principal designer role is usually seen at larger companies that have surpassed an initial stage of major growth.

At smaller companies, teams generally have a clear mandate regarding the required investment. This renders a principal designer unnecessary. While an individual might be hired into this level at small and medium companies for compensation reasons, existing employees may find it challenging to organically grow to this level—the organization hasn't yet reached the level of complexity at which they would benefit from a principal designer.

Similar to the transition from the senior to staff level, the growth trajectory from the staff to principal level is usually lengthy. The level of responsibility at the principal level requires serious credibility and trust. Demonstrating consistency at that level takes time. Designers can expect to spend three or more years at the staff level before being promoted to a principal level.

Nico Matson, Senior Principal Product Designer at Yahoo, has had a lengthy design career spanning 16 years. At Yahoo, her title is the equivalent to the level above staff. She spent four years at the staff level before landing her current position. "They look at us to shape the product direction and help leaders decide what business strategies to lean into." Nico stated that everything else about the role is similar to staff, but *more*:

- Crafting a clear and direct story that's compelling to any type of audience
- Working cross-functionally with partners on a strategic and tactical level
- Building up the team's craft and giving them guidance on how to get to this level

Due to the high level of ambiguity that comes with the role, individuals who want to grow from staff to principal must build a proficiency in defining their own workload and directing their energy. While a staff designer likely aligns with their manager about projects they take on, many principal designers operate more autonomously. They often seek out and validate potential investments the company might benefit from, ultimately using design to make the maximum business impact.

At the principal level and above, the altitude of work is stratospheric. Principal and above designers are highly versatile problem-solvers that can be thrown at any situation and self-orient. One week, they might help the executive team develop the company's vision for a product or a portfolio of products. The next week, they might be pitching an idea for a new large-scale initiative.

Because these individuals are usually so removed from the tactical efforts of product teams, they must work hard to maintain an understanding of lower-level work. Principal designers often appear in design critiques on a rotating basis or sit in design reviews alongside other members of senior leadership. They may also offer office hours for a similar purpose. In these situations, they usually provide hands-on feedback about design implementation and offer support in situations where lower-level designers feel squeezed.

> **TIP INDUSTRY-WIDE IMPACT**
>
> In all ways, principal designers are more prolific than staff designers. Many, but not all, designers at this level also make an impact on the industry at large. Jina Anne, Principal Product Designer at Microsoft, is the creator of the design tokens concept that powers modern design systems and organizes Clarity, a design systems conference. Val Head, Principal Product Designer at Adobe, is on the board of the W3C (World Wide Web Consortium), speaks about CSS animations at conferences worldwide, runs a podcast about animation, and has published a book on the subject, *Designing Interface Animation*.
>
> That said, many principal and above designers have little to no online presence, do not participate in conferences, and refrain from public speaking or writing. Instead, they direct their energy into impact at their employer. Both kinds of impact matter, so every designer is free to choose their preference.

Switch Roles

While some designers continue to hone their craft and ascend the career ladder to principal and beyond, many realize their interests lie elsewhere. These individuals instead decide to make lateral moves. There are many adjacent roles that a designer's skills can apply to, but the most common transitions include:

- Content Design and Information Architecture
- Design Operations

- Design Management
- Product Management
- User Research
- UX Engineering

Most designers interact with folks in these roles on a regular basis, so they may seem familiar. Designers who find any of the following descriptions of these positions interesting may be able to either try a temporary shift at their current employer or take a class to explore responsibilities of the role before diving in. Designers can also consult folks in their network who currently occupy the position they are interested in by scheduling an exploratory coffee chat. People are often happy to share their thought process and experiences with others!

CATT'S CORNER

Irritation Is Natural

I've had many conversations with designers in which they wonder if design is right for them. In fact, I've asked myself this exact question over and over again. As part of the process of figuring out my career direction, I regularly speak with people in adjacent roles to see if those might be a better fit for me. But I keep coming back to design as the fit for my set of skills.

I've come to the realization that there will just be times when design work is irritating or unfulfilling. It's easy to be annoyed by a job when you've been doing it for long enough to see so many of its downsides. Designers are asked to do so many different kinds of work, and we take on so much emotional labor. It's normal to be overwhelmed and disillusioned.

Even so, when I think about why I got into design, the rationale still resonates with me. I still feel magical when I illustrate potential futures. And I still feel proud when someone uses an experience I designed—especially when they appreciate it.

Through my own introspection, I learned that I need to reconnect regularly with the reasons I'm a designer. Whenever possible, I seek out design activities that naturally give me energy. When I discuss potential work options with my manager, I remind them of what invigorates me and what drains me. This helps us work together to design a workload that gives me more connection to my purpose.

If you find yourself feeling particularly disillusioned by design work, I recommend you consider the root cause of the issue. What got you into design in the first place? What gives you energy? In *Designing Your Life*, Bill Burnett and Dave Evans recommend tracking your work life and highlighting the energy boosts versus drains (Figure 9.7). This can help you advocate for a better situation.

Date	Time	Activity	Engagement	Energy	Flow	Notes
10/30/2019	11:00 AM	Design work	10	4	Yes	worked in Figma on some new design ideas
10/30/2019	12:00 PM	1:1	6	2	No	met an old coworker for coffee and talked about design operations
10/30/2019	12:45 PM	Team lunch	4	-3		awkward team lunch
10/30/2019	1:30 PM	Planning meeting	5	-3		planning for team trip to Hudson. not sure how folks feel about it
10/30/2019	2:00 PM	Planning meeting	8	0		more planning with less people. it was clear who needed to be in the room. much more productive and hopeful
10/30/2019	2:30 PM	Design work	8	4		more design work in preparation for share-out
10/30/2019	4:00 PM	Sync	8	4	Yes	showed my work to the person who will use it and she was super excited!
10/30/2019	5:00 PM	Planning meeting	8	4		Team training overview. it was wonderful to feel a sense of hope!

FIGURE 9.7

In 2019, I documented my energy boosts and drains to identify the situations in which I felt most aligned with my role.

Sometimes, the root cause is that your current situation at work is specifically burning you out. If you used to enjoy design work but find it more tedious lately, consider if your environment is the issue. Perhaps it's time to switch teams, change employers, or take on some freelance work.

If you're truly ready to shift away from design, that's also okay! Design work is best suited for people who are excited by the practice, as it involves creativity. If you're mentally blocked due to lack of connection to design overall, you'll struggle to create the innovative solutions that users and businesses may require. In that case, I hope the following descriptions and interviews can give you some insight into common pros and cons of each role. ■

Content Design and Information Architecture

Similar to UX and product designers, a content designer focuses on the structure of information within an end-to-end customer experience. However, these individuals are generally more concerned with the words and objects an experience is composed of. On average, content designers usually don't focus on visual design at all. Instead, they own the hierarchy of information and decide where it appears. At some companies, this role may be more similar to a UX design position. At others, it might align more with UX writing. Every company treats content design differently.

IN THE REAL WORLD

NICHOLE MILLER-KREZELAK

Nichole Miller-Krezelak is currently a Content Designer. She worked as a writer and content strategist for 15 years and went to design school to become a UI/UX Designer in 2019. Nichole spent two years doing full-stack product design work, and then naturally transitioned to content design when she accepted a role with the title at a start-up. "From there, I continued to get recruited for content design roles."

As a content designer, Nichole now leverages insights to craft empathetic, intuitive user experiences through the lens of information hierarchy and terminologies. She is satisfied with the shift in name because her work outcomes are still the same—"I am first and foremost a UX designer. Whether I'm designing with words or pushing pixels, I approach design holistically. Great design needs to be a balance of perfect UI interfaces met with on-tone, of-that-moment language."

Designers who shift to content design should be aware that many companies consider this role to be extraneous. Because many of the skills overlap with other kinds of designers, a content design practice can feel like a luxury that most small start-ups cannot afford. Content designers at many organizations spend part of their time advocating for the value of their role. This trade-off may or may not be worth the risks depending on how much you love words or despise visual design.

Design Operations

Design program management and operations center around the logistical side of design work. This change would be ideal for individuals who find themselves energized by the idea of improving operation's efficiency. Design operations folks focus on tooling and process, ensuring that their design team has all the resources they need to perform predictably and effectively. They are essentially designing the experience of being a designer, creating the environment a designer needs to succeed.

IN THE REAL WORLD

TAMAR PACHECO-THEARD

Tamar Pacheco-Theard began her career as a Graphic and Web Designer in 2012. By 2016, she was a Senior UX Designer at Rakuten. Over time, she became "much less interested in retail, much more interested in employee experience." Tamar focused her energy on visualizing "how I might be able to be the change I wished to see in the world through design ops," which led her to found the design operations discipline at Rakuten. She then joined Asana in 2021 as a Senior Design Program Manager on the design operations team. Since then, she has expanded her scope from design to the broader engineering, product, and design group.

The design operations function is most common at larger organizations where the maintenance of a team's tools and workflows becomes a full-time job. A team of 30 or more designers requires proper attention, or individuals will start going rogue and working in ways that are not conducive to creating great experiences. That said, this position is also viewed as nice-to-have because it results in an increase of efficiency—i.e., a reduction in costs—rather than an increase in key metrics.

Design Management

A design manager is responsible for building a team of talented individuals and directing their energy so they may be successful. This role is familiar to most designers because they have likely reported to a design manager before. This job is best suited for people who are excited by the idea of helping designers grow as people. It involves a lot of emotional labor and requires grit and self-awareness. Therefore, designers who are considering this shift must invest heavily in their emotional intelligence or EQ. Sometimes, design managers have to share hard news or let go of reports, and it's important to be mentally prepared for the emotional strain that can come from that kind of communication work.

While companies generally value design managers, this role is also sometimes at risk due to the perception that more management layers result in less efficiency. There is a constant pendulum swing between the idea that there are too many management layers and too many direct reports for one manager to handle. This shift may result in a manager having only one report, followed by nine or more.

Some managers have needed to shift back to hands-on design to keep their employment. Consider the pros and cons as much as possible and be open to the likelihood that you will have some great moments and some frustrating moments when you step into a design manager position.

IN THE REAL WORLD

KRISTEN LEACH

Kristen Leach first entered the industry as a Designer in 2013. She continued to level up and eventually reached the senior level during her time as a Product Designer at Etsy. In 2020, she left Etsy and led the design for the digital version of an independent game called Wavelength.

While wrapping up designs for the Wavelength app, Kristen joined a financial tech company as a Product Design Lead. She transitioned to a manager role within her first year. The position was offered to her due to a critical business need: "They said, 'Kristen, can you do this? Because no one else can do this.'"

Kristen agreed and was so adept at the role that she was eventually tapped to lead an entire group. Her favorite parts of design management are high-altitude strategic direction and building a talented team. "There are people who are just better at the hands-on work than me. I love to let them shine."

Solving people problems gives Kristen the same joy, if not more, as she gets from navigating design challenges. "Designers are very thoughtful, empathetic people who care about others." Since making the switch to management, she has continued to deepen her management practice. She took on a founding designer role for a couple of years at a gaming start-up and then shifted into a director role at a major eCommerce platform.

Product Management

Designers who love strategy might be interested in diving head-first into the world of business by becoming a product manager. Many designers have made the shift into PM—it's one of the most common transitions in the industry. Product managers make the final call regarding a project's scope, and many designers get so frustrated by their product managers that they decide to take the reins themselves. Some designers fill in for their PM already, writing documentation and making scope decisions independently, so they decide to get credit for their work by making the title change official.

Designers who are curious about product management must learn as much as they can about the way that businesses work. Product management is more than just writing docs; it's all about risk analysis and deduction. Check out *Product Management for UX People* by Christian Crumlish for a strong introduction to the world of product management. Designers don't need business degrees to make the transition, but they should be able to speak the language and meet expectations of the role.

IN THE REAL WORLD

CAMBRIA KLINE

Cambria Kline began her career as a Graphic Designer in 2008, eventually shifting into web design in 2010 before becoming a Product Designer in 2012. She continued to advance in her product design career and eventually became a Senior Product Designer at a major eCommerce marketplace in 2016.

Cambria didn't feel like product design was the right fit for her strengths. "Instead of focusing solely on becoming a better designer, I kept my scope broad, building skills that typically fell under my cross-functional partners' roles." This effort helped her collaborate more effectively with her teammates and gave her insight into where her interests lay beyond design.

continues

IN THE REAL WORLD (continued)

In 2015, Cambria had the opportunity to operate as an Interim Product Manager while the team backfilled a vacant role. "It started with curiosity about my product peer's work." Balancing product management with her design workload helped her learn more about the position without officially making the switch. The vacant role was filled after three months, so she then returned to hands-on design work for another year before experimenting with another potential career direction: design management. While it felt like a step forward, the design management role still didn't connect with the type of responsibilities that fueled her energy. So, when the opportunity presented itself in 2018, she moved into a full-time product management role.

Cambria feels like her product design experience helped her to become a successful PM. "I'm still involved in the parts of design I enjoyed—digging into user needs, brainstorming solutions, and testing ideas creatively before fully committing to a vision." She believes it also helps her to be a stronger partner to her teammates. "I see my partners and leadership team as my users and use my design background to communicate ideas clearly, identify opportunities to improve team processes, and collaborate on solutions for better efficiency."

Cambria excelled in her new role—she grew from a mid-level to a staff-level PM in the span of three years! But she recommends that designers focus on what matters most to them rather than getting caught up in the pursuit of a title. "We don't have to commit to one path forever, and even if we do, that path will likely evolve."

User Research

Designers who enjoy connecting with customers and conducting research sessions might just be primed for a transition into full-time UX research work. This is an especially simple position to shift into for many designers who already have experience conducting UX research sessions. If you enjoy finding new research methods and discovering more ways to gather qualitative insight but don't find joy in defining the customer experience itself, UX research might be for you.

IN THE REAL WORLD

ROSE KUE

Rose Kue first became a UX designer in 2015 after transitioning from a prior career spanning 10 years. By 2016, she was promoted to a Senior Designer role. After a year, she shifted to user experience research internally. "It aligned better with my past experience, and I found it more interesting," she said. "Design required more pattern recognition and matching." This wasn't how Rose wanted to spend her time.

Instead, Rose wanted to spend time "observing, learning, and changing the narrative about how colleagues think about problems and solutions." User research was a natural fit for her. She has continued to invest in her research practice and is now a Senior Staff Researcher at a health tech company.

Designers who are very opinionated about the way an interface is organized and visually treated may want to shy away from going into user research full-time, as they will no longer be able to make design decisions but rather only influence them. A common complaint that researchers make is that teams don't integrate their recommendations often enough. Anyone who shifts from design to research must be willing to completely let go of how their design uses insights and instead fully embody their role as an advisor.

UX Engineering

Designers who can code in HTML, CSS, and JavaScript might find UX Engineering or Creative Technologist roles to be of interest. UX engineers partner with designers and engineers to implement customer experiences so that they are the highest level of quality possible. Because they have a design background, they can add polish that other engineers might ignore. These kinds of engineers are especially valuable on design systems teams and can also make a major impact in situations where noncode prototypes aren't helping the team determine next steps.

IN THE REAL WORLD

ADEKUNLE ODUYE

Adekunle Oduye began his career as a web designer and developer in 2011 and then became a Product Designer in 2014. He continued to work as a Product Designer for four years; he then took on a Design Technologist role. In this role, he combined his design skill set with his programming knowledge to test, build, and maintain various UI components for the Memorial Sloan Kettering Cancer Center's design system. This led him down the path of UX engineering.

"My transition from Product Designer to UX Engineer was driven by my desire to take concepts from idea to production. As my development skills grew, I became better at shaping the overall UX and validating ideas early, which ensures that my colleagues and I don't waste time building the wrong solution." Over the four years since his transition to UX engineering, Adekunle has grown to a more senior level in this career path. He has worked in this capacity with companies such as Plaid and Mailchimp.

UX engineers are still relatively rare. There have been front-end engineers for years now, but the shift from front-end to UX has taken some time to clarify. UX engineers often feel like they are in between worlds—some are managed by design leaders, while others are managed by engineering leaders. Neither design nor engineering usually knows how to utilize their strengths so they end up doing a lot of advocacy work. If you want to make the shift, prepare to explain the benefits of the role often.

That's All, Folks!

Design careers are complex and last longer than ever before. The needs of customers also get more complicated every day. There will be no shortage of work for those who continue to sharpen their critical thinking skills and refine their communication abilities.

Whether you decide to dig further into the world of staff design by investing in the skills covered in this book, or you decide to shift into an adjacent role, or leave the industry altogether, either way, please remember to do three things:

- Give yourself grace.
- Rest often.
- Follow the fun!

Give Yourself Grace

Like any form of growth, professional development takes time. No one can grow without making mistakes. Have compassion for yourself and allow yourself to be imperfect. When you accept your faults, you begin to see opportunities for change. Make sure to be nice to yourself and make space for expected moments of cringe along your journey. Just like design, lots of other skills also require consistent practice to see improvement and everyone grows at a different pace. You can't compare yourself directly to others because they had their own path and you have yours.

Rest Often

Design is a creative practice. Like other forms of creativity, designers sometimes reach a point of burnout. It's important to build a consistent practice of rest so you can bring your sharpest mind to work.

Dr. Saundra Dalton-Smith came up with a concept known as the seven types of rest. As shown in Figure 9.8, the types of rest include creative, emotional, mental, physical, sensory, social, and spiritual. Most designers specifically need mental and creative rest.

According to Dr. Dalton-Smith, mental rest is needed when a person is feeling "irritable and forgetful" and struggles to turn off their brains "as conversations from the day fill [their] thoughts." Individuals in need of mental rest need time for their brain to reset and recharge. Examples of mental rest activities include taking a break to go for a device-free walk, going on a vacation away from work, and meditation. Creating rules about taking time off and incorporating daily walks or other forms of exercise into your schedule can help you ensure that you have regular and consistent moments to take a brain break so your brain doesn't burn out.

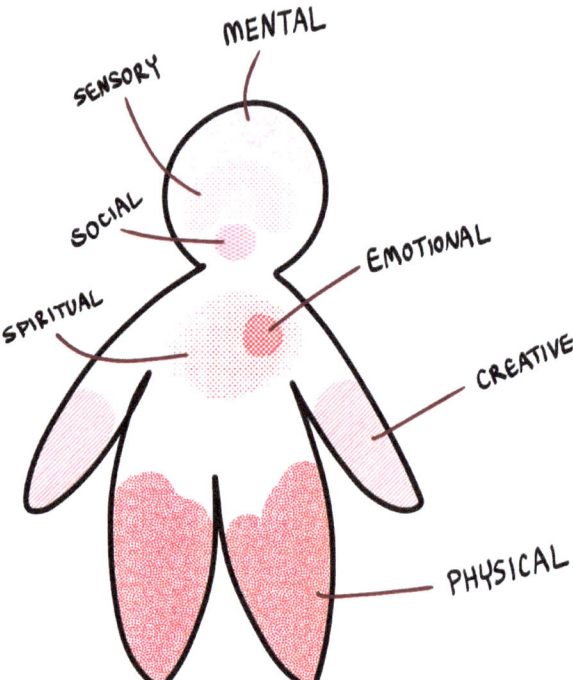

FIGURE 9.8
The seven types of rest, created by Dr. Saundra Dalton-Smith.

Creative rest is "important for anyone who must solve problems or brainstorm new ideas," wrote Dr. Dalton-Smith in a piece for TED's *How to Be a Better Human* series, titled "The 7 Types of Rest That Every Person Needs." Considering the expectations of product design and UX work, designers must regularly ensure that they get creative rest so they can consistently innovate. If you ever feel like you're hitting a wall with your design work, it might be time to take some creative rest. Go to a museum and observe some art. Take a workshop in the fine arts at a local small business or continuing education department of a university. Allow yourself to create and be inspired by others with no expectations of quality.

Some designers also require emotional and spiritual rest. If you don't have space to rest, you will become emotionally exhausted by the amount of patience and labor required to move from idea to implementation of a design. This is why supporters are so crucial—some of them will be able to validate your feelings and give you the space you need to feel whole. Many designers are driven by purpose and need to feel like their work contributes to something bigger. If you are needing this but it's not coming from work, you may need to connect to your local community or find a hobby that can give you that sense of purpose.

Follow the Fun

While design work *is* work, it's also a creative practice. Creativity demands curiosity. As you learn more about what gives you energy, take note and dig deeper. Uncover more questions with every answer and never stop learning.

Try things of all kinds, even things that don't seem directly related to design. You never know how something you learn might connect to your work as a designer. Inspiration comes in all forms, and some of the best designers are multihyphenates.

Debrief

There are so many directions you can choose to go with your career. This ambiguity might feel scary, but it also means you are free. You can keep ascending the career ladder as an individual contributor or you can go into a different field. You can try lots of new hobbies. Or you can look for a new job. The world is your oyster—search for the pearls. Lean into the discomfort, give yourself grace, and have fun!

INDEX

A

actions to claim agency as designers, 106–111
    fidelity and upscaling progressively, 106, 109–110
    insight and confidence, 106, 107–108
    power and curiosity, 106–107
    silos and designing in public, 106, 110–111
    table, leaving a seat at. *See* seat at the table
active voice, 175–176
activities, at end of chapters
    building influence, 141–144
    delegation, 164–166
    design org, 47–48
    managing your presence, 188–190
    power relationships, 98
    staff designer archetypes, 27–28
    vision creation, 119–120
    workload capacity, 75–76
additive business impact, 135–136
affinity map of symptoms, 96
akimbo pose, 173–174
ambiguity navigation, for hiring team interviews, 194
analytics, 132
Anne, Jina, 199
annotated designs, as presentation illustrations, 138
Any.do personal task management tool, 61
appeal of a rejection, 139
approvers, in your network, 78–79, 84–85
archetypes of staff designers, 20–26, 27
    activity, 27–28
    architect, 20, 21
    platformer, 20, 25–26
    tastemaker, 20, 22–23
    visionary, 20, 23–24
architect, as archetype of staff designers, 20, 21, 27
artifacts, in presentations, 138, 144
Asana
    staff designer position at, 41
    vision for future of work, 100
    visionary illustration from, 116
    work management tool, 65
asynchronous communication methods, 179–180
asynchronous presentation, 138–139
automation tools, 54

B

backlogs, 64–65, 147
battles, choosing, 126–129
"before-and-after" examples, 186
Bennett, Micah, 124
Blanchard, Ken, 158
blogs, as resources for career growth, 37
body language, 173–174
boundary setting, 66–74
    assessing tasks before acceptance, 70
    estimating capacity, 67–69
    saying "no," 72–74
    staggering projects to buy time, 71–72
Budiu, Raluca, 108
Buley, Leah, 108
Burnett, Bill, 200

burnout, 50, 66, 125, 140, 209
business impact
    additive, subtractive, and qualitative, 135–136
    for hiring team interviews, 196–197
business strategy
    designers as facilitators of, 101–102
    as focus area for designers, 9–10, 21, 22, 23
    as overlap between staff designers and managers, 16, 18
    prioritize problems to build influence, 128

C

calendar
    blocking off focus time, 51–52
    coding blocks of time with color and emojis, 59–60
    time management tools, 53–55
candor, in managing your presence, 176–177
capacity. *See* workload capacity
career paths beyond staff design, 191–211
    changing companies, 193–197. *See also* interviewing at new company
    directions to go, 192
    grace, rest, and fun, 208–211
    rise beyond staff, 197–199
    switch roles, 199–208. *See also* roles in companies
career paths for designers, 2–3, 4–7, 37
Catt's Corner
    buy yourself time, 71–72
    communications and power relationships, 92
    designing in public, 110–111
    disillusionment, and energy boosts vs. drains, 200–201
    influencing org design challenges, 127–128
    mentorship, 156–157
    pacing and perfectionism, 45–46
    sharing failures, 184–185
    workloads and altitudes, 15–16
Chen, Lil, 41, 52, 79
churn, in re-orgs, 30, 154
Clarity design systems conference, 199
co-creation, 158–159
collaboration with others, for hiring team interviews, 195
color-coding blocks of time on calendar, 59–60
communication
    active voice vs. passive voice, 175–176
    being curious rather than critical, 107
    of boundaries, 73–74
    contextual with power relationships, 92
    frequencies, 180–181
    methods: asynchronous and synchronous, 179–180
    mistakes in speaking, 176
    of value of staff designer, 168
content design role, 201–202
context, as information in updates for approvers, 181–183
context switching, 56–58
contextual communications with power relationships, 92
contractor, as external delegation, 154–155
contributors, in your network, 78–79, 82, 83
control of influence. *See* Locus of Control
craft
    as focus area for designers, 9–10, 23, 25
    as overlap between staff designers and managers, 16, 17–18
    quality, sharing for inspiration, 157
creative rest, 210
creative technologist role, 207–208
creativity, curiosity, and fun, 211
credit, giving its due, 93, 187
Crumlish, Christian, 205

curiosity
  at beginning of new project, 106–107
  and creativity and fun, 211
customer feedback, for hiring team interviews, 196

D

Dale, Jess, 180
Dalton-Smith, Saundra, 209–210
day in life of staff designer, 11–13
decision, as information in updates for approvers, 181–182
delegation, 145–166
  activity, 164–166
  categories of work to delegate, 149–151
  choosing how to, 151
  external delegation, 151, 154–155
  facilitating vs. micromanaging, 156–163. *See also* facilitating vs. micromanaging in delegation
  handling challenges, 162–163
  internal delegation, 151, 152–154
  knowing when to delegate, 146–147
  the right work to the right people, 147
delivery, in sharing your case, 138–139
demeanor, in managing your presence, 173–174
design career ladder, 4–7. *See also* career paths
design guidelines, 160–161
design management, 203–204
design managers, compared with staff designers, 16–18
design operations, 202–203
design tokens concept, 199
Design Week, on designers' seat at the table, 101
"Designers Have a Seat at the Top Table—So What Now?" (Lord), 101
Designing Interface Animation (Head), 199

Designing the Gap (Martin), 17
Designing Your Life (Burnett and Evans), 200
diagrams, in presentations, 138
directness
  and candor, 176–177
  of direction, 159–161
disappointment, handling, 139–140
documentation, internal, for organizational insights, 131
documentation of findings, 86–97
  dynamics (relationships of power), 87–92
  opportunities (new initiatives), 92–93
  sentiment (people's feelings), 93–97
  symptoms (stories), 97
Dominguez, Nicole, 33
Double Diamond, 68–69
dynamics (relationships of power), 86–92
  high power, 87–88, 92
  low power, 89, 92
  moderate power, 88, 92
  power mapping, 89–91

E

emoji coding on calendars, 59–60
emotional rest, 210
emotions, in narrative for vision projects, 114
energy boosts vs. drains, 200–201
estimating design work, 67–69
Evans, Dave, 200
events, as resources for career growth, 37
examples
  of backlog tasks, 64
  "before-and-after," 186
  of design work, sharing, 161
  of projects of staff designer, 15–16
  of questions. *See* questions, examples of

of ways to communicate your boundaries, 73–74
of ways to unblock engineers, 72
executive presence, 172–173
exits, for medium companies, 38
expectations
    setting, in managing up, 177, 178–181
    of staff designer, 8–10
exploratory conversations, 130–131, 200
external delegation, 151, 154–155
    permanent employee, 155
    temporary/freelance/contractor, 154–155

F

Fabrigar, Leandre R., 114
facilitating vs. micromanaging in delegation, 156–163
    co-creating, 158–159
    craft quality, 157
    directness with direction, 159–162
    positive reinforcement, 162
    situational leadership, 157–158
    thought processes, 158
failures, sharing, 184–185
Fantastical scheduling tool, 53
fidelity of visual design matching fidelity of thinking
    in presentations, 137–138
    in product vision, 102, 103–104, 109–110, 116–117
Figueroa, Sarrah, 34, 105, 175
five whys, 184–185
focus areas for designers, 8–10
focus time
    blocking off on calendar, 51–52
    and context switching, 56–58
foundation of presentation, 136–137
founders, at small companies, 33–36
freelance employee, as external delegation, 154–155

G

glue work, 129
Google, staff designer position at, 41
grace, giving yourself, 209
Gretz, Alison, 127

H

Harllee, Jessica, xx, 80
Harvard Business Review (HBR), on executive presence, 172
Head, Val, 199
Head of Design, 31, 39, 42
Hersey, Paul, 158
high-fidelity concepting, 103–104
high-power people, 87–88, 92
HiPPOs (highest paid person's opinion), 85, 112
hiring teams. *See* interviewing at new company
Hoe, Asia, 127
Hogan, Lara, 59
Huff, Jason, 43–44, 117
humility, 170–171
Hunt, Randy, 104

I

ideation, and fidelity mismatch, 103–104
illustrations
    in creation of a vision, 100, 103, 109–110, 116–117
    in presentations, 137–138
imposter syndrome, 171
individual contributors (ICs), 3, 122, 127–128
industry-wide impact, 199
influence, 121–144
    activity, 141–144
    build your case: observation, proposal, outcome, 132–136, 143
    defined, 122, 124
    gather research, 130–132

influence (*continued*)
  handling rejections, 139–140
  how it works, 122–126
  Locus of Control, 122–123, 125–126
  pick your battles, 126–129
  share your case, 136–139
information architecture role, 202
insights. *See* research
internal delegation, 151, 152–154
  partnership, 152–153
  reorganization, 153–154
  rotation, 153
internal research, 130–131
interviewing at new company, for career change, 193–197
  ambiguity navigation, 194
  business impact, 196–197
  collaboration with others, 195
  critical project work, 193
  customer feedback, 196
  portfolio presentations, 193–194
Irabor, Iyobosa, 89, 122

J

Jira project management tool, 62, 65
Johnson, Prarthana, 186
journey maps, in presentations, 138

K

keeping receipts, 140
Kline, Cambria, 205–206
Kue, Rose, 207
Kushwaha, Kritika, 170
Kwon, Michelle, 32, 159

L

Lambridis, Scott, 101
language. *See also* communication
  and succinctness, active voice vs. passive voice, 175–176

large companies, 41–46
lateral moves, in career path. *See* roles in companies
Leach, Kristen, 204
lead designers, 5
leadership techniques, 157–162
Lee, Justine, 58
Levitt, Debbie, 101
Lin, Elizabeth, 23
Locus of Control, for influence, 122–123, 125–126
Logan, Meghan, 40–41, 107, 126
Lord, Gemma, 101
Lovin, Brian, xviii–xix, xxi
low-power people, 89, 92
lunch, in work day, 11

M

The Making of a Manager (Zhuo), xx
management career track, 2
management layers in companies, 31, 37
management roles
  design management, 203–204
  product management, 205–206
managers, compared with staff designers, 16–18
managing up, 177–187
  project completion announcement, 184–187
  regular updates, 177, 181–183
  setting expectations, 177, 178–181
managing your presence, 172–177
  candor, 176–177
  demeanor, 173–174
  succinctness, 175–176
Martin, Ben, 17
Matson, Nico, 198
medium companies, 36, 38–41
meetings, auditing in advance, 58–59
Mendelow, Aubrey L., 91
mental rest, 209–210

mentorship
  of designers, being a good mentor, 156–157, 159
  as overlap between staff designers and managers, 16–17
Merholz, Peter, 17–18
metrics, 132, 134, 179, 196–197
micromanaging. *See* facilitating vs. micromanaging in delegation
Miller-Krezelak, Nichole, 202
miscellaneous tasks, in backlogs, 64–65
mock-ups, as presentation illustrations, 137
moderate-power people, 88, 92
Morris, Edwin, 45, 97
Motion automation tool, 54
moving on, from a rejection, 140

N

narrative, for vision projects, 114–116
Natoli, Joe, 108
negative sentiments, 94, 96–97
network building, 78–86
  approvers, 78–79, 84–85
  contributors, 78–79, 82, 83
  observers, 78–79, 82, 84
  supporters, 78–79, 85–86
neutral sentiments, 94
"The New Rules of Executive Presence" (HBR), 172–173
"no," saying it more, 72–73
notetaking tools, 81
Notion work management tool, 63
numerical data, 132

O

observation, in building a case, 133–134
observers, in your network, 78–79, 82, 84
Oduye, Adekunle, 208
offboarding of temporary employees, 155
Ohama, Yoko Sakao, 7, 112

onboarding
  in internal and external delegation, 153, 154, 155
  at medium companies, 40, 151
  project brief, 66
opportunities (new initiatives), 86–87, 92–93
O'Reilly, Charles, 30
org design, impact of, 29–48
  activity, 47–48
  defined, 30
  effect on success, 30–31
  influencing challenges, 127–128
  large companies, 41–46
  medium companies, 36, 38–41
  navigating challenges, 43–44
  small companies (start-ups), 33–36, 37
Org Design for Design Orgs (Merholz), 18
organizational insights, 130–131
outcomes
  in building a case, 133, 135–136
  in project completion announcement, 184, 186
output of work, in project completion announcement, 184, 186
ownership areas, and picking battles, 129

P

Pacheco-Theard, Tamar, 203
pair design sessions, 159
partnership, as internal delegation method, 152–153
passive voice, 175–176
Pearce, Jen, 113, 149
pen and paper notes, 81
perception management, 172
perfectionism, 45
perfectionism-driven silos, 102, 104, 110
permanent employee, as external delegation, 155
personal task management tools, 61, 63
personas, 108, 138

persuasion toolkit, 122
Petty, Richard E., 114
planning projects, 65–66
platformer, as archetype of staff designers, 20, 25–26, 27
player-coach work, 5
podcasts, as resources for career growth, 37
portfolio presentations, for hiring team interviews, 193–197
positive reinforcement, 162
positive sentiments, 93–94
power in relationships. *See* dynamics (relationships of power)
power in vision work
    curiosity and questions, 106–107
    renounced, 102
power mapping, 89–91
power ranking, 89–90
presence. *See* managing your presence
presentation of your case, 136–138
presentations, synchronous and asynchronous, in delivery, 138–139
principal designer, as career beyond staff, 197–199
principal title, 5–6
prioritization matrix, 70
process, in project completion announcement, 184, 187
product design, 4
product management, 205–206
Product Management for UX People (Crumlish), 205
product vision, 111–112. *See also* vision work
professional development. *See* career paths
project briefs, 65–66, 179
project completion announcement, 184–187
project management tools, 62–63
project planning, 65–66

project work, for hiring team interviews, 193
projects, staggering to buy time, 71–72
promotions, 7
proposal, in building a case, 133, 134–135
prototypes, as presentation illustrations, 138
"The Psychology of Design," 108
pushback, as delegation challenge, 162
Pyton, Kristina, 73

Q

qualitative business impact, 136
qualitative user research
    for building influence, 131–132
    insight for vision work, 102, 103, 107–108
    scarcity at small and medium companies, 40–41
quantitative user research, for building influence, 132
questions, examples of
    for collaborators, 80
    of curiosity, at beginning of new project, 106–107
    when gathering research, 130

R

Radical Candor (Scott), xx
re-orgs. *See* reorganization
real world, designers in the
    Adekunle Oduye, 208
    Angira Shirahatti, 19
    Cambria Kline, 205–206
    Fiona Rolander, 169
    Jessica Harllee, 80. *See also* Harllee, Jessica
    Kristen Leach, 204
    Lil Chen, 52. *See also* Chen, Lil
    Micah Bennett, 124

Michelle Kwon, 32. *See also* Kwon, Michelle
Nichole Miller-Krezelak, 202
Rose Kue, 207
Sarrah Figueroa, 105. *See also* Figueroa, Sarrah
Sunnie Sang, 148. *See also* Sang, Sunnie
Tamar Pacheco-Theard, 203
Yoko Sakao Ohama, 7. *See also* Ohama, Yoko Sakao
"The Rebalancing of Design Management" (Watkins), 18
Reilly, Tanya, 129
rejection, handling, 139–140
relationships, 77–98
  activity, 98
  approvers, 78–79, 84–85
  contributors, 78–79, 82, 83
  as core principle, xxi, xxii
  documentation of insights, 86–87
  dynamics, 86–92. *See also* dynamics (relationships of power)
  influence of people, 125
  network building, 78–79
  observers, 78–79, 82, 84
  opportunities (new initiatives), 86–87, 92–93
  sentiment, 86–87, 93–97. *See also* sentiment (people's feelings)
  supporters, 78–79, 85–86
  symptoms (stories), 86–87, 97
renounced power, 102
reorganization (re-org)
  defined, 30
  as internal delegation method, 153–154
research
  building influence with insights, 130–132
  at small and medium companies, 40–41
  for vision work, 102, 103, 107–108
resources
  for career growth, 37
  as information in updates for approvers, 181–183
rest, seven types of, 209–210
Restivo, Anthony, 113–114
return on investment (ROI), 186
Rolander, Fiona, 169
roles in companies, 199–208
  content design, 201–202
  design management, 203–204
  design operations, 202–203
  information architecture, 202
  product management, 205–206
  user research, 206–207
  UX engineering, 207–208
rotation, as internal delegation method, 153
Rotter, Julian B., 122–123

S

Sang, Sunnie, 51–52, 70, 148
scale
  to build influence, 132
  as core principle, xxi, xxii
  effect on speed, 45
scheduling tools, 53
scope, for vision projects, 113–114, 119
seat at the table, 101–105
  designers as facilitators of business strategy, 101–102
  fidelity mismatch, 102, 103–104
  lack of insight, 102, 103
  missed opportunities, 102
  perfectionism-driven silos, 102, 104
  renounced power, 102
  taking back the wheel. *See* actions to claim agency as designers
self-promotion, 170–171
senior designers, compared with staff designers, 14–16

sentiment map, 95
sentiment (people's feelings), 86–87, 93–97
    negative, 94, 96–97
    neutral, 94
    positive, 93–94
"The 7 Types of Rest That Every Person Needs" (Dalton-Smith), 210
Shedroff, Nathan, 101
Shirahatti, Angira, 19
silos in vision work, 102, 104, 110–111
situational leadership, 157–158
sketches,
    as presentation illustrations, 137
    hand-drawn, 109
Slate, Theresa, 100
slideshows, as presentations, 137
Small Business Administration, on company sizes, 33
small companies (start-ups), 33–36, 37
Smith, Matt D. (MDS), 23
social capital, 126–127, 139
software engineering career track, 2
spiritual rest, 210
staff designer, defined, 1–28
    archetypes, 20–26, 27–28
    career ladder for, 2–3, 4–7
    a day in the life of, 11–13
    expectations of, 8–10
    not a manager, 10, 16–18
    not a senior designer, 10, 14–16
    role of, 2–3
staff engineering career track, 2
The Staff Engineer's Path (Reilly), 129
staff title, 4
stakeholder matrix, 89, 91
start-ups, small companies as, 33
storytelling, by visionaries, 23, 24
strategy. *See* business strategy
Stripe, org design of, 30
subtractive business impact, 136

succinctness, in managing your presence, 175–176
super-senior designers, xxi, 3
supporters, in your network, 78–79, 85–86
sustainability, as core principle, xxi, xxii
"swoop and poop," 158–159
symptom affinity map, 96
symptoms (stories), 86–87, 97
synchronous communication methods, 179–180
synchronous presentation, 138–139
systems thinking, as focus area for designers, 9, 21, 23, 25

T

Takigayama, Tom, 197
Tan, Yan Ling, 127
target audience, 115
tastemaker, as archetype of staff designers, 20, 22–23, 27
team project management tools, 62–63
temporary employee, as external delegation, 154–155
thought processes, sharing, 158
time and workload capacity, 49–76
    activity, 75–76
    capturing the workload, 60–66. *See also* workload capacity
    definitions of time and workload, 50–51
    making time for visioning, 51–60. *See also* time for visioning
    setting boundaries, 66–74. *See also* boundary setting
time crunches, as delegation challenge, 162–163
time for visioning, 51–60
    calendar blocks, color-codes and emojis, 59–60
    calendar blocks for focus time, 51–52
    context switching, 56–58

defined, 50
meeting audits, 58–59
time management tools, 53–55
timeline for creation of a vision, 114
Timely tracking tool, 55
Todoist task management tool, 65
Toyoda, Sakichi, 184
tracking tools
   for time management, 55
   for work management, 61–63, 65
Turman, Kyle, 117
Tushman, Michael, 30

U

unicorn companies, 38
updates, regular, in managing up, 177, 181–183
The User Experience Team of One (Buley and Natoli), 108
user research. *See* qualitative user research; quantitative user research
user research role, 206–207
UX engineering, 207–208

V

value of staff designers, 167–190
   activity, 188–190
   communication of, with intrinsic and extrinsic methods, 168
   humility, 170–171
   managing up, 177–187
   managing your presence, 172–177
vision, value of, 111–118
   creation of a vision, 113–117, 120
   product vision, 111–112
   sharing the vision, and distribution of, 117–118
   visioning, defined, 111
   when to propose a vision, 112–113
vision types, as presentation illustrations, 138

vision work, 99–120
   activity, 119–120
   a seat at the table, 101–105. *See also* seat at the table
   taking action, 106–111. *See also* actions to claim agency as designers
   value of a vision, 111–118. *See also* vision, value of
visionary, as archetype of staff designers, 20, 23–24, 27
visioning. *See also* time for visioning
   defined, 111
visual and interaction design skills, 9
visual design
   and fidelity of thinking, 103–104, 109–110, 116–117
   as focus area for designers, 9–10, 22
   illustrations in presentations, 137–138

W

Watkins, Cap, 18
Wegener, Duane T., 114
wireframes, as presentation illustrations, 137
work management tools, 61–63, 65
working agreements, 83
working manual, 83
workload capacity, 60–66
   backlog and miscellaneous work, 64–65
   defined, 50–51
   estimation of, 67–69
   project planning, 65–66
   work management tools, 61–63
workshops, for organizational insights, 131
write-ups, as presentations, 136–137

Y

Yue, Maggie, 115

ACKNOWLEDGMENTS

There are so many people I would like to thank for making this book possible. I wish I had infinite pages to list every person I've spoken to about this book. Since I have limited space, I will try to be concise with my gratitude. Here we go!

My community of creative comrades, especially: Asia Hoe, Christine Ryu, Erin Nolan, Jen Pearce, Jess Dale, Jessica Harllee, Kristen Leach, Lil Chen, Micah Bennett, Michelle Kwon, Namika Hamasaki, Rox Ravago, Sarah Ohye, Shanique Shields, Stephanie Lawrence, Von Chan, and Yoko Sakao Ohama.

Prolific thinkers who trusted me with their time and words: Adekunle Oduye, Anthony Restivo, Brian Lovin, Cambria Kline, Edwin Morris, Fiona Rolander, Jason Huff, Kritika Kushwaha, Kyle Turman, Maggie Yue, Meghan Logan, Nichole Miller-Krezelak, Nico Matson, Prarthana Johnson, Randy Hunt, Rose Kue, Sarrah Figueroa, Sunnie Sang, Theresa Slate, and Tom Takigayama.

Everyone who gave feedback on various drafts: Aria Todd, Ariel Cotton, Brian Carr, Brian Chung, Casey Tang, Cat Lo, Cheryl Platz, Guo Chen, Hang Xu, Izzy Oji, Jakub Rybar, Jess Dale, Jon Tsay, José Arias, Julie Xie, Kim Alban, Kristine Sanchez Henson, Liz Osaki, Michael Carruthers, Nikisha Roberts, Parker Simon, Senongo Akpem, and Treyce Meredith.

Additional industry friends and collaborators who inspire me to be better: Abby Covert, Adelle Lin, Alexandra Brown, Amy Lima, Anna Sternoff, Camille Acey, Carlos Montoya, Cate Huston, Chris Meeks, Femke Van Schoonhoven, Gilbert Ghang, Harrison Wheeler, Jenny Wen, Jina Anne, Joel Califa, La Vesha Parker, Linda Eliasen, Mahsino Blamoh, Maurice Cherry, May-Li Khoe, Mike Wandelmeier, Mina Markham, Rachel Hsiung, Reginé Gilbert, Sabrina Hall, Sandra Bilbrey, Sara Kremer, Sara Wachter-Boettcher, Senongo Akpem, and Skye Zhang.

Friends and family who kept me grounded when I was stress-melting during the writing process: Mom, Dad, Aileen Zhou, Helena Wentworth, Laura Donohue, Natalie Liebert, giftcrabs, and the LaG '07 Misfits.

The Rosenfeld Media team, more specifically these fine folks for all of their support: Julia Hansen, Karen Corbett, Louis Rosenfeld, Marta Justak, and Zontee Hou.

Finally, I want to thank all 155 members of the first four cohorts of my Staff Designer course. Y'all helped me go from a figment of an idea to a curriculum to a manuscript. I am infinitely grateful for every conversation we had together.

This book only exists because of all of you. Thank you. We did it!

 Rosenfeld

Dear Reader,

Thanks very much for purchasing this book. There's a story behind it and every product we create at Rosenfeld Media.

Since the early 1990s, I've been a User Experience consultant, conference presenter, workshop instructor, and author. (I'm probably best-known for having cowritten *Information Architecture for the Web and Beyond*.) In each of these roles, I've been frustrated by the missed opportunities to apply UX principles and practices.

I started Rosenfeld Media in 2005 with the goal of publishing books whose design and development showed that a publisher could practice what it preached. Since then, we've expanded into producing industry-leading conferences and workshops. In all cases, UX has helped us create better, more successful products—just as you would expect. From employing user research to drive the design of our books and conference programs, to working closely with our conference speakers on their talks, to caring deeply about customer service, we practice what we preach every day.

Please visit rosenfeldmedia.com to learn more about our **conferences**, **workshops**, **free communities**, and **other great resources** that we've made for you. And send your ideas, suggestions, and concerns my way: louis@rosenfeldmedia.com

I'd love to hear from you, and I hope you enjoy the book!

Lou Rosenfeld,
Publisher